ENVIRONMENTAL SCIENCE, ENGINEERING AND TECHNOLOGY SERIES

RECENT PROGRESS ON EARTHQUAKE GEOLOGY

Environmental Science, Engineering and Technology Series

Nitrous Oxide Emissions Research Progress
Adam I. Sheldon and Edward P. Barnhart (Editors)
2009. ISBN: 978-1-60692-267-5

Fundamentals and Applications of Biosorption Isotherms, Kinetics and Thermodynamics
Yu Liu and Jianlong Wang (Editors)
2009. ISBN: 978-1-60741-169-7

Environmental Effects of Off-Highway Vehicles
Douglas S. Ouren, Christopher Haas, Cynthia P. Melcher, Susan C. Stewart, Phadrea D. Ponds, Natalie R. Sexton Lucy Burris, Tammy Fancher and Zachary H. Bowen
2009. ISBN: 978-1-60692-936-0

Agricultural Runoff, Coastal Engineering and Flooding
Christopher A. Hudspeth and Timothy E. Reeve (Editors)
2009. ISBN: 978-1-60741-097-3

Agricultural Runoff, Coastal Engineering and Flooding
Christopher A. Hudspeth and Timothy E. Reeve (Editors)
2009. ISBN: 978-1-60876-608-6
(Online Book)

Conservation of Natural Resources
Nikolas J. Kudrow (Editor)
2009. ISBN: 978-1-60741-178-9

Conservation of Natural Resources
Nikolas J. Kudrow (Editor)
2009. ISBN: 978-1-60876-642-6
(Online Book)

Directory of Conservation Funding Sources for Developing Countries: Conservation Biology, Education and Training, Fellowships and Scholarships
Alfred O. Owino and Joseph O. Oyugi
2009. ISBN: 978-1-60741-367-7

Forest Canopies: Forest Production, Ecosystem Health and Climate Conditions
Jason D. Creighton and Paul J. Roney (Editors)
2009. ISBN: 978-1-60741-457-5

Soil Fertility
Derek P. Lucero and Joseph E. Boggs (Editors)
2009. ISBN: 978-1-60741-466-7

Handbook of Environmental Policy
Johannes Meijer and Arjan der Berg (Editors)
2009. ISBN: 978-1-60741-635-7

The Amazon Gold Rush and Environmental Mercury Contamination
Daniel Marcos Bonotto and Ene Glória da Silveira
2009. ISBN: 978-1-60741-609-8

Process Engineering in Plant-Based Products
Hongzhang Chen
2009. ISBN: 978-1-60741-962-4

Buildings and the Environment
Jonas Nemecek and Patrik Schulz (Editors)
2009. ISBN: 978-1-60876-128-9

Tree Growth: Influences, Layers and Types
Wesley P. Karam (Editor)
2009. ISBN: 978-1-60741-784-2

Syngas: Production Methods, Post Treatment and Economics
Adorjan Kurucz and Izsak Bencik (Editors)
2009. ISBN: 978-1-60741-841-2

Potential of Activated Sludge Utilization
Xiaoyi Yang
2010. ISBN: 978-1-60876-019-0

Psychological Approaches to Sustainability: Current Trends in Theory, Research and Applications
Victor Corral-Verdugo, Cirilo H. Garcia-Cadena and Martha Frias-Armenta (Editors)
2010. ISBN: 978-1-60876-356-6

Carbon Capture and Storage including Coal-Fired Power Plants
Todd P. Carington (Editor)
2009. ISBN: 978-1-60741-196-3

Process Engineering in Plant-Based Products
Hongzhang Chen
2009. ISBN: 978-1-60741-962-4

Fluid Waste Disposal
Kay W. Canton (Editor)
2010. ISBN: 978-1-60741-915-0

Recent Progress on Earthquake Geology
Pierpaolo Guarnieri (Editor)
2010. ISBN: 978-1-60876-147-0

ENVIRONMENTAL SCIENCE, ENGINEERING AND TECHNOLOGY SERIES

RECENT PROGRESS ON EARTHQUAKE GEOLOGY

PIERPAOLO GUARNIERI
EDITOR

Nova Science Publishers, Inc.
New York

For permission to use material from this book please contact us:
Telephone 631-231-7269; Fax 631-231-8175
Web Site: http://www.novapublishers.com

Library of Congress Cataloging-in-Publication Data

Recent progress on earthquake geology / editor, Pierpaolo Guarnieri.
p. cm.
Includes bibliographical references and index.
ISBN 978-1-60876-147-0 (hardcover)
1. Earthquakes. I. Guarnieri, Pierpaolo.
QE534.3.R43 2009
551.22--dc22

2009049574

Published by Nova Science Publishers, Inc. ✢ New York

CONTENTS

PREFACE

This book includes papers of researchers from different countries around the world that point out the study of seismoinduced phenomena associated with recent and historical earthquakes, along with an article about the April 6, 2009 devastating earthquake in L'Aquila in central Italy.

Starting from basic knowledge, it is possible to reconstruct the geo-history of the phenomena induced by earthquakes and their recourrence time. Earthquake geologists study the surface expression of earthquakes, such as faults and ground failure caused by strong ground shaking. Moreover, ground deformation phenomena and other secondary effects, such as tectonic subsidence and uplift, liquefaction, shaking-induced landslides, tsunamis and the migration of hydrogen as protons can be used to identify earthquakes. Finally, with paleoseismology—a branch of earthquake geology—we can define timing, location and size of prehistoric earthquakes.

The collected articles highlight the importance of the geology of the events—in other words, all of the information recorded in the stratigraphic levels. In fact, only through the reconstruction of the past history can we try to forecast future events; otherwise, we can only attempt prevention. In fact, the knowledge of when, where, how often, and with what magnitude large earthquakes occur is crucial for understanding and characterizing the seismic hazard of a region. For this purpose, the chapters regard the recent progress in the study of recent and past seismic events on the basis of the descriptions and analysis of primary effects (surface ruptures) and secondary ones (tsunamis and liquefactions).

Chapter 1 – On the April 6, 2009, a Mw 6.3 earthquake struck the town of L'Aquila in central Italy. It was followed by a long seismic crisis with another five events with Mw between 5.1 and 5.6. Available seismological data point out an extensional deformation field with an average WSW-ENE tensional axis. In the course of the seismic sequence, two distinct én échelon fault sources were activated: first, the SW-dipping Paganica normal fault, which is associated with the Mw 6.3 event; and, subsequently, the southern part of the WSW-dipping Gorzano normal fault.

Co-seismic ground deformation (open fissures, én-échelon cracks, minor faults within the fault damage-zone, and rejuvenated fault scarps with centimetric downthrows) have been surveyed for a length of nearly 13 km along the Paganica fault.

The integration of the information from this last Italian earthquake with the previous seismotectonic background has allowed us to further detail the 3-D shape and the size of some of the individual seismogenic sources of the Apennine active extensional belt.

Chapter 2 - Among the seismogenic paleorift basins around the world, the Kachchh rift perhaps occupies a unique position. This Mesozoic rift located in northwest India, remarkable for its higher seismogenic productivity, is the only intraplate region where two large earthquakes have occurred in a period less than 200 years. This region has been largely unexplored since the first historically known earthquake in 1819, but the recent advances in paleoseismological research have facilitated reconstruction of its eventful past. Results of studies during the last decade have exposed a seismically active past, with multiple locations associated with varying recurrence intervals. The 1819 earthquake located on the northern flanks of the Kachachh rift that resulted in a spectacular landform (Allah Bund) was contrasted by the 2001 earthquake, which occurred on a blind thrust, within the rift system. Both these earthquakes have produced seismically induced liquefaction features in a large area, and these have served as proxies to older features. The authors' studies in these regions exposed liquefaction features ~1000 and 4000 years old, which seem to be predecessors of the 1819 and 2001 events, respectively. At least one other independent source, which seems to have been activated during the last 2000 years, was identified through our investigations. Evidence for still older events, with less certain source zones, have also been documented. This chapter presents an evaluation of the paleoseismologic investigations in the Kachchh region, leading to the estimation of timing of past events.

Chapter 3 - Calabria, the southern tip of the Italian boot, is characterized by some of the strongest earthquakes of the whole Mediterranean area. Seismicity mainly clusters between the seventeenth and the beginning of the twentieth century, with major events exceeding Magnitude 7. This region is also affected by different tectonic domains, as can be argued by the distribution of the seismicity itself, as well as by the cinematic characteristics of the instrumental seismicity and by the studies on active tectonics.

There is general agreement about the active extension within the back-arc sector (i.e., the Tyrrhenian side). On the contrary, no agreement exists on the seismogenic behavior of the central part of the Calabrian Arc and, mainly, of its external zone (Ionian side).

This study focuses on an area in the eastern sector of the northern Calabria Arc, contributing to understand the relationships between some historical earthquakes that occurred here (i.e., the 1836 A.D., the 1824 A.D., and the ~ 951 A.D. earthquakes) and their causative fault(s). The studied area underwent, since the Quaternary, a strong regional uplift of about 1 mm/y and is expected to be characterized by a generalized compressive remote stress field, as is typical of the frontal parts of the orogenic belts. Some geological and geomorphological evidence suggests that, as such a compressive tectonic regime is quite recent; however, field survey demonstrated that E-W striking normal faults are very young structures in the area, extending the crust in an N-S direction, as also shown by previous GPS data.

Archive research studies were used to re-evaluate the distribution of the effects for the last seismic event that occurred in the area (the M>6, 1836 earthquake). Even if the intensity distribution is spread in a wide area lengthened for about 70 km, the highest values cluster along the E-W striking Rossano fault. This fault, showing a spectacular scarp, strongly controls the morphology of the area and has been investigated in this study with a geo-structural and palaeo-seismological approach. The very young activity of the fault (post 125 ka) is suggested by the geomorphological analysis of marine and alluvional terraces widely exposed in the area. Besides, an explorative paleoseismological trench allowed dating the age of the colluvial wedges associated with the fault, then contributing to constrain the date of the last fault re-activation, which fits with the 1836 A.D. earthquake.

All these observations allowed re-locating the epicentre of the 1836 A.D. earthquake along the Rossano fault, which can be considered as the main on shore located active structure of the area.

The field observations were used to attempt a parameterization of the fault (slip rate, time recurrence interval, maximum magnitude expected) and have strong implication for the seismic hazard of the Ionian side of the northern Calabria.

Chapter 4 - To learn about recurrence of large earthquakes and strain model of the Düzce segment of the seismogenic North Anatolian fault zone (November 1999, M 7.1), systematic and independent geomorphological and paleoseismological approaches were integrated and compared to overcome intrinsic limitations of both methods. Offset geomorphic markers allowed measurements of cumulative slip during repeated events and estimates of the fault slip rate over middle- and long-term periods (7000–60,000 years), with no information on the timing and patterns of faulting events. Conversely, investigation by paleoseismologic trenching provided a small number of paleoearthquakes that cover short-term periods (<2,000 years).

Geomorphological mapping along the fault trace permitted reconstruction and analysis of offset geomorphic markers. Faulted fluvial terraces and deflections of stream crossing the fault trace were useful for cumulative deformation estimates. By means of OSL and radiocarbon dating methods, the onset of the offset were dated Late Pleistocene and Early Holocene, yielding a rate of deformation of 15.0±3.2 mm/yr for the last 60 kyr. On the basis of a "characteristic-earthquake model" and under constant slip rate assumptions, the stream deflections were estimated to be produced by repetition of 20–30 1999-like earthquakes during the last 7000 years, suggesting 290±60 years of recurrence time for surface rupturing events.

By merging information obtained from paleoseismological trenching, evidence for five surface faulting events were found and, on the basis of radiocarbon, 210Pb and archaeological dating, the paleoearthquake record for the past two millennia was reconstructed. This record is suggestive of bimodal recurrence distribution yielding overall average recurrence time of ~350 yrs, compatible with that calculated from the geomorphic markers analysis. Under the assumption of "characteristic earthquake", bimodal paleoearthquakes distribution indicate slip rate fluctuations during the past two millennia, with clustered high strain release. These results suggest a Wallace-type strain release model for the Düzce fault, and average strain accumulation of 11.4±1.2 mm/yr, in agreement with slip rate results for the past 60 kyr obtained by geomorphic marker analysis.

Chapter 5 - The study of paleotsunami deposits is a new line of research only recently undertaken in Italy but also in the broad Mediterranean. Therefore, approaches to apply for the identification and dating of tsunami deposits need severe testing, systematization and formalization. This work was carried out in eastern Sicily following a multi-theme approach and testing several different methodologies. Interesting and promising results came from an original combination of geomorphological, paleontological, X-ray, petro-chemical, morphoscopic and magnetic analyses.

Eastern Sicily has been affected in historical times by large local tsunamigenic earthquakes, such as the 1169, 1693 and 1908 events and by tsunamis originating in the eastern Mediterranean. Starting from information on the effects of known tsunamis combined with a geomorphological approach, several sites potentially suitable for preserving tsunami deposits, such as coastal lakes, marshes and lagoons, were selected. In these sites, 64 test gouge cores have been dug by hand and engine coring in order to reconstruct paleoenvironments and to identify potential paleotsunami deposits.

The deposits interpreted as tsunami-related rarely share the same characteristics, which depend substantially on the depositional environment, on the local geomorphology and the type of the beach and offshore deposits. In general, the tsunami deposits represent layers different from the normal stratigraphic sequence, often coarse-grained chaotic units as result of high-energy depositional events. In some cases, micropaleontological analysis confirmed the marine origin of these anomalous units. Their ages are constrained both with radiocarbon dating on charcoal fragments or shells and tephra analysis. Dating of the tsunami layers and their correlation with known historical tsunamis represented a critical issue for the final results of tsunami deposit studies.

At Capo Peloro in north-eastern Sicily, combining archaeological, historical, and C14 data, two tsunami deposits have been associated to the 1783 A.D. and 17 earthquakes. Evidence for the occurrence of multiple inundations in the eastern flank of Mt. Etna sites has been also collected: three events in the past 580 yrs at Anguillara site and four events in the past 4000 yrs at Gurna site. In southeastern Sicily, in the Augusta bay, combining historical, tephrostratigraphical and C14 dating, a tsunami inundation history composed of six events in the past 4000 yrs has been reconstructed. The two most recent tsunami deposits are related to the 1693 and 1169 earthquakes, whereas one is probably linked to the 365 A.D. Crete event and one to the 3500 B.P. Santorini eruption-triggered tsunami.

Chapter 6 - The Apulian foreland flexure below the Calabrian and Hellenic Chains generated a system of normal faults, mainly NNW-SSE oriented, during the Pliocene-Pleistocene Age. These are clearly evident on seismic reflection profiles, and are generally interpreted as the present main tectonic features of the foreland. A detailed analysis of the available data have instead highlighted, along the western margin and in the South Salento offshore, the presence of several recent compressive features which have been assigned to inversion tectonics, probably developing since the Middle Pleistocene Age. The present regime would be the result of compressive/transpressive stress domain, as testified by several earthquakes, transferred to the deep Apulian crust by the Calabrian and Hellenic Chains. The fault blocks, previously tilted during the Upper Pliocene/Lower Pleistocene Age, have successively back rotated by inversion tectonics, so to produce the uplift of the rigid blocks, the seismic events recorded in the region and the compressive deformation, especially evident inner the upper soft sediments. In this way, the morphology of the seismic area of the South Salento offshore is characterized by gentle surface swells, typically placed above the hanging wall of the post-Miocene extensive faults. The present SSE-ward migration of the Calabrian Arc, parallel to the western Apulian margin, transferred also the important strike-slip component to the deep level of the foreland.

Chapter 7 - The essential contribution of structural geology in the identification and parameterization of individual seismogenic sources is now largely recognised. When both geological and good-quality instrumental seismological data are available, the 3D fault geometry and segmentation pattern, the kinematics (sense of slip and slip rate) and the associated stress tensor can be defined in detail. The identification of the 3D geometry of historically activated and/or silent sources, which is fundamental for seismic hazard assessment, is evidently more difficult. It may strongly benefit from the combined help of structural geology and regional tectonics, which may allow constraining a long-term geologic and kinematic context for the present deformation field and associated seismicity.

The close integration of geological and seismological data has allowed us to identify a seismogenic extensional fault system which extends for about 500 km across northern-central Italy from Lunigiana, in northern Tuscany, to Val di Sangro in southern Abruzzo. The fault system, which mainly consists of SW-dipping high-angle normal faults developed within the hanging wall of an east-dipping low-angle detachment system, has been named Etrurian Fault System.

The authors' approach to the identification and parameterisation of the associated 3D individual seismogenic sources has been structural-seismotectonic: they have integrated surface geology data (trace of active faults, i.e., 2D features) with subsurface data (depth distribution of aftershock zones, background seismicity, active seismological data, crustal rheology; i.e., a 3D approach). Geological observations have allowed them to define a segmentation model consisting of major fault structures separated by first-order (kilometric scale) structural-geometric complexities considered as likely barriers to the propagation of major earthquake ruptures. The authors have named "seismogenic boxes" the plan projection of such major active structures and for each of them the maximum magnitude of the expected earthquake can be evaluated. From northern Tuscany to southern Abruzzo, we have identified and parameterized 37 boxes that are responsible or capable of experiencing earthquakes with $M \geq 5.5$. These boxes lie within the boundary of the EFS extensional seismogenic province.

The boxes account for almost all the largest earthquakes known in the region (I>VIII, $M \geq 5.5$), but they do not contain all the minor seismicity. The latter may be enclosed within wider zones, of well-defined tectonic setting and seismotectonic meaning, derived from a plausible seismotectonic model of the region. The seismotectonic model we used provides that the upper crust seismicity is extensional and mainly confined within the hanging wall of an east-dipping low-angle normal fault zone. The surface trace of the EFS separates a western crustal sector, only slightly seismic, from an eastern highly seismic crustal sector which is undergoing active SW-NE extension at the EFS hanging wall. The synthetic and antithetic faults within the EFS hanging wall subdivide this extending rock volume into minor structural blocks which correspond to seismotectonic zones at the surface.

The structural-seismotectonic approach is suitable for a seismotectonic zonation able to improve the seismic hazard assessment. In particular, the three-dimensional reconstruction of the seismogenic master fault geometry should represent a strong constraint to delimit the zones within which the largest earthquakes are expected to occur.

Chapter 8 - Proposed results of research are based on the conception of proton tectogenesis, occurring owing to the migration of hydrogen as protons from the Earth's core into near-earth outer space. The model for a hydride Earth core was met with support of research in which the "intumescence" of iron metal samples in fluid hydrogen under superhigh pressure has been reported. The "intumescence" of iron metal samples saturated by hydrogen alone explains the Earth core density decrease in comparison with "pure iron" theories.

Båth's world-wide seismo-effect is connected with the proton tectogenesis conception. Such seismo-effect led to conception of world-wide seismo-resource, cumulation of which supports a decrease of regional seismo-resources in the remaining parts of world's subterranean except the zone of gigantic earthquake and its aftershocks. Such seismo-effect is evidence of world-wide links between earthquakes foci situated thousands kilometers apart. The operationability of those links eliminates their explanation with the help of mechanical models and forces the acceptance of an electric grid model for interpretation of a seismo-globalism event.

The research of a world-wide seismic process has given particular attention to cosmic-meteo-seismo correlations. For the first time, B. Gutenberg pointed to these correlations in paragraph "§11 Parallelismus zwischen Erdbebenhäufigkeit und meteorologischen Elementen sowie kosmischen Erscheinungen" in his 1927 book. H. Alfven discussed the electric grid nature of interactions of different parts of solar-terrestrial space. For example, he considered connections between solar wind, magnetosphere and the ionosphere of the Earth. The whole process of world-wide proton tectogenesis is considered in the framework of the electric grid model. Protons, as positive elementary charges, support with operationability the world-wide connections between earthquake focuses. In the research of a world-wide seismic process, it was found necessary to give particular attention to electric phenomena observed before, during and after earthquakes.

On the basis of the concept of proton tectogenesis, the possibility of taking measurements of electric impulses of proton migration with the help of vertical sequence of electrodes in the earth was discovered.

The proposed results show differences between electromagnetic interpretation of results of ULF magnetic field measurements and interpretation of measurements of subterranean EMF in the framework of current-networks.

Special attention is given to the results of subterranean electric measurements on Kamchatka for illustration of world-wide seismic processes, specifically for the tsunamigenic earthquake that occurred in Indonesia (2004 December 26) and the damaging earthquake in China (2008 May 12).

— Pierpaolo Guarnieri
Department of Geological Sciences, University of Catania, Corso Italia, 55 95129 Catania, Italy

Now at GEUS (The Geological Survey of Denmark and Greenland), Øster Voldgade, 10 1350 Copenhagen, Denmark

In: Recent Progress on Earthquake Geology
Editor: Pierpaolo Guarnieri, pp. 1-17
ISBN: 978-1-60876-147-0

Chapter 1

THE APRIL 2009 L'AQUILA (CENTRAL ITALY) SEISMIC SEQUENCE (M_W 6.3): A PRELIMINARY SEISMOTECTONIC PICTURE

***Giusy Lavecchia**[*1], **Paolo Boncio**[1], **Francesco Brozzetti**[1], **Rita de Nardis**[2], **Deborah di Naccio**[1], **Federica Ferrarini**[1], **Alberto Pizzi**[1] and **Giuseppe Pomposo**[1]*

[1] Dipartimento di Scienze della Terra,
Univ. "G. d'Annunzio" – Campus Univ., 66013 Chieti, Italy
[2] Dipartimento della Protezione Civile, Rome, Italy

ABSTRACT

On April 6, 2009, an M_w 6.3 earthquake struck the town of L'Aquila in the Abruzzo region in central Italy. It was followed by a long seismic crisis with other four events with M_w between 5.1 and 5.6. Seismological and geological data point out an upper crust extensional stress field with an average WSW-ENE tensional axis. In the course of the seismic sequence, two distinct en échelon fault sources were activated: first, the SW-dipping Paganica normal fault, which is associated with the M_w 6.3 event; and, subsequently, the southern part of the WSW-dipping Gorzano normal fault.

* Corresponding author: glavecchia@unich.it

Co-seismic ground deformation (open fissures, en échelon cracks and shear planes with centimetric downthrows) was surveyed for ~ 13 km along the Paganica fault. The integration of the information from this last Italian earthquake with the previous seismotectonic background has allowed us to further detail the 3-D shape and the size of some of the individual seismogenic sources of the Apennine active extensional belt.

Keywords: earthquake, seismogenic source, normal fault, fault segmentation, L'Aquila, Apennines, central Italy

INTRODUCTION

On Monday, April 6, 2009, an M_w 6.3 earthquake struck the town of L'Aquila (Abruzzo region) in the core of the central Apennine mountain range, bringing death (about 300 casualties) and destruction. It did not arrive unexpectedly; still, we were not prepared for it.

The Abruzzo Apennines and the L'Aquila area have a well-documented and long seismic history. During the second half of twelfth century, under the Aragon domination, L'Aquila become an important commercial and artistic centre, due to a flourishing economy founded on saffron trading. Unfortunately, it experienced two destructive earthquakes in 1349 (M_w 6.5) and 1461 (M_w 6.5) and several minor ones in the following two centuries. However, the heavier seismic period occurred in the eighteenth century, with five moderate-to-large earthquakes (1703, M_w 6.7; 1750, M_w 5.0; 1762, M_w 5.9; 1786 M_w 5.2; 1791, M_w 5.4), the most damaging one occurring in February 1703 and producing extensive destruction in L'Aquila, with thousands victims. Collapse of monuments and city walls were widespread, as well as aqueduct breakages and surface land effects. The nineteenth and twentieth centuries were spotted with moderate events, with magnitude generally lower than M_w 5.5 (Working Group CPTI04, 2004). This relatively long time of non-destructive earthquakes determined a general underestimation of the seismic risk among the local authorities. However, the Earth Science Department at the "G.d'Annunzio" University of Chieti produced a number of papers and research studies on the seismotectonics of central Italy, strongly based on the field evidence of active faulting and focused on seismic hazard assessment (Lavecchia et al., 2002; Pizzi et al., 2002; Boncio et al., 2004a; Boncio et al., 2004b; Pace et al., 2006; Brozzetti et al., 2009; Pizzi and Galadini, 2009). Furthermore, an extensive report on the seismic hazard in the Abruzzo region was drawn up by Lavecchia and co-authors and given to the local political

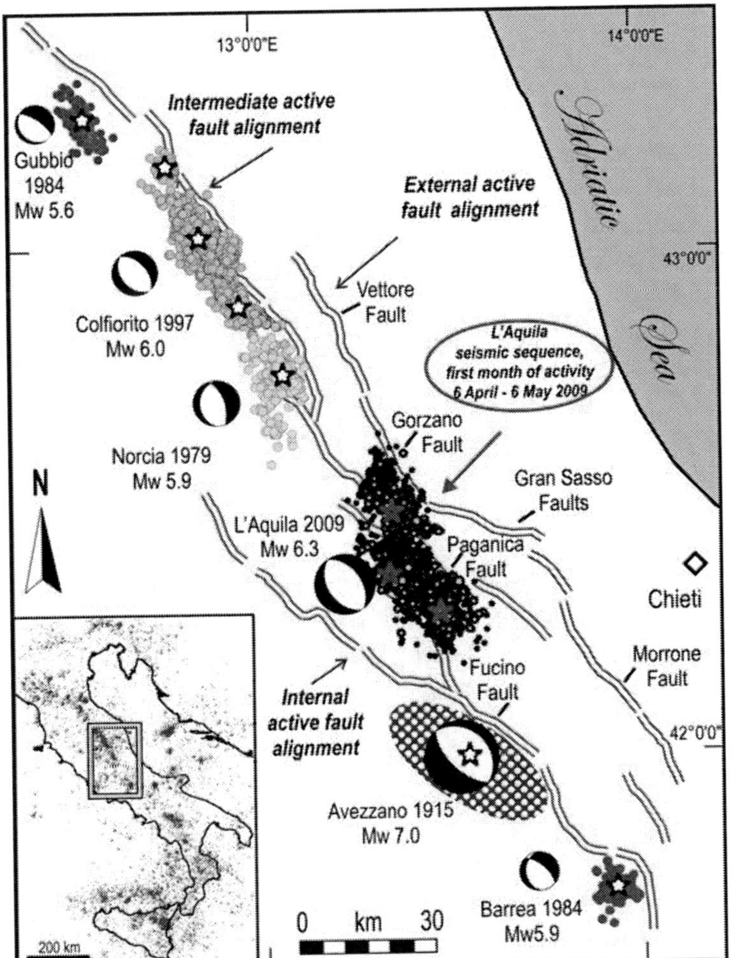

Figure 1. Location map of the L'Aquila 2009 earthquake (updated to May 6) in comparison with the epicentral distribution and the focal mechanisms of the major ($M_w \geq$ 5.5) instrumental seismic sequences of the Umbria-Marche-Apennines (central Italy) in the last thirty years; the stars indicate the epicentral location of the major events, the three red ones being the major events of the L'Aquila sequence (April 6 M_w 6.3, April 7 M_w 5.6, April 9 M_w 5.4). The epicentral area of the Avezzano 1915 earthquake is schematically reported together with its focal mechanism (Ward and Valensise, 1989). Major, active and possibly seismogenic, normal and normal-oblique fault systems, extending with an average NNW-SSE strike across the Apennine mountain belt, are schematized into three distinct alignments, named internal, intermediate and external (see Lavecchia et al., this book). The crustal seismic activity ($h \leq 30$ km) recorded across the entire Italian territory in the January 1981–January 2009 time frame, is reported in the inset.

authorities in 2006, with the aim of raise the issue of large, potential seismic sources in a heavily populated area. The current earthquake was not a surprise, and its main features and distribution fit well with our previous seismic hazard evaluations.

At the time we are writing this chapter (beginning of May 2009), the L'Aquila seismic sequence is still ongoing and we can only present a preliminary description of it and of some related surface effects, but a discussion of the earthquake in the framework of previous knowledge about the seismotectonics and the individual seismogenic sources of the Apennines can be started. We are confident that our interpretation will be improved and refined, but not substantially changed, when the high-quality seismological data, presently collected and processed by the INGV (Istituto Nazionale di Geofisica e Vulcanologia) researchers, will be made available for the broad scientific community. Nevertheless, the present earthquake can add some more information to the general seismotectonic picture given in another chapter of this book by Lavecchia et al., which we wrote before the earthquake. The earthquake has allowed us to focus on some aspects, previously unknown, and to fill some knowledge gaps concerning the segmentation of the active faults and their behaviour during the earthquake propagation.

THE SEISMIC SEQUENCE

The main shock (Ml 5.8, M_w 6.3) of the L'Aquila seismic sequence occurred on the night of April 6, at 01:32 UTC. It was preceded by some foreshocks in the previous days (three events with $3 \leq M_w < 4$) and was followed by a long period of seismic activity in the following weeks. As of today (May 6, 2009), just one month after the main shock, the sequence is still active and is progressively growing NNW-ward (www.ingv.it).

From March 30 to May 6, an area of nearly 600 km^2 extending for almost 45 km along an average NNW-SSE direction (Figure 1), has been struck by thousands of seismic events, including more than 150 Ml $\geq$ 3.0 (according to the Italian Seismic Bulletin database by INGV), with a corresponding total energy release of about 1.5E+14J. This value is about half an order of magnitude greater than the energy released during the last strong seismic sequence in central Italy (September–December 1997 Colfiorito earthquake, M_w 6.0). The seismic activity associated with the L'Aquila earthquake involved three, initially separated and now interconnected, epicentral areas, here named L'Aquila, Campotosto and Ocre (Figure 2).

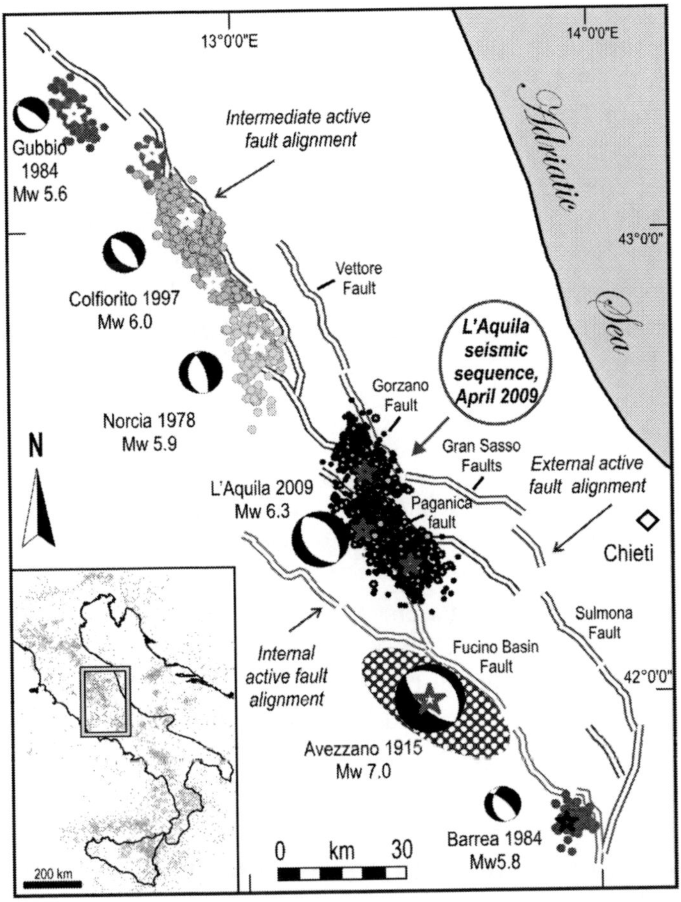

Figure 2. Epicentral distribution of the L'Aquila 2009 seismic sequence from March 30 to May 6 with focal mechanisms of the three foreshock events with $3 \leq M_w < 4$, of the three major events (M_w 6.3, 5.6 and 5.4) and of the other events with $M_w \geq 4.0$. The overall seismological dataset is available on line (*www.ingv.it*). The focal mechanisms are divided into three geographically distinct groups (L'Aquila, Ocre and Campotosto) and for each of them an average focal mechanism was computed through the application of the Binghan statistics; the earthquake time of occurrence is indicated as day/month/year. The common stress tensor, computed through application of the Carey inversion method (Carey-Gailhardis and Mercier, 1987) is shown in the stereonet (Wulff projection) of the right side corner; the great circles with arrows represent the nodal planes with related slip vectors; the histogram gives the angular divergence between the theoretical (t) and observed (s) slip vectors; R is the stress ratio $(\sigma_2 - \sigma_1)/(\sigma_3 - \sigma_1)$. The tensional axes (T) of the average focal mechanisms and the main stress axes (σ_1, σ_2, σ_3) are given as trend (first 3 numbers) and plunge (last 2 numbers).

The L'Aquila area was the first to be activated by the foreshock events from March 30 to April 5, by the main event, with M_w 6.3, on April 6 and by three other events with $4<M_w<5$ in the first two days; the hypocentral depths of all these events range from 7 to 10 km, the main shock being located at a depth of 9.5 km. The Ocre area, located some kilometres southeast-ward, was significantly activated since April 7; it was struck by one event with M_w 5.6 and by three events with $4.0<M_w<4.5$; these four major events are located at depth values between 8 and 15 km. The Campotosto area, located northward of the main shock epicentral area, was struck by two relatively strong events (M_w 5.4 and 5.3) on April 9, by another $M_w>5.0$ earthquake on April 13, and by four more events with $4<M_w<5$; the hypocentral depths of these major events range from 8 to 11 km.

The focal mechanisms of the major events ($M_w\geq4$) of the L'Aquila seismic sequence have a common, almost pure, extensional kinematics (Figure 2). We divided them into three separated groups, corresponding to the three geographically distinct epicentral areas, and for each group the average fault plane solution was calculated. The corresponding T-axis has a WSW-ENE trend for the L'Aquila and Campotosto areas and a SW-NE trend for the Ocre area. In order to verify the compatibility of all the mechanisms with a common stress tensor, we inverted the data set using the Carey-Gailhardis and Mercier (1987) method. All the mechanisms have resulted to be compatible with a common tensional stress tensor, whose representative ellipsoid is triaxial (shape factor $R = (\sigma_2-\sigma_1)/(\sigma_3-\sigma_1) = 0.33$), with a horizontal N54°E-trending σ_3 axis (Figure 2).

ACTIVE FAULTING IN THE EPICENTRAL AREA

The epicentral area of the L'Aquila seismic sequence crosses two distinct NNW-SSE-striking alignments of active normal faults (Figure 1). The western system develops from Gubbio to Colfiorito, Norcia, Cittareale, Pizzoli, L'Aquila, Paganica and Navelli; the eastern system develops from Mt. Vettore to Mt. Gorzano and Gran Sasso and still southwards to the Sulmona area. Another extensional system, not involved in the seismic activity this time, extends from Valle del Salto to Magnola, the Fucino Basin and Barrea.

The three alignments, named respectively internal, external and intermediate (Figure 1, Lavecchia et al., this book and reference therein), offset Mio-Pliocene fold-and-thrust structures and generate intramontane depressions. Displacements of Late Pleistocene-Holocene deposits and morphologies, as well as paleoseismic activity are widely documented (Galadini and Galli, 2000; Valensise and Pantosti,

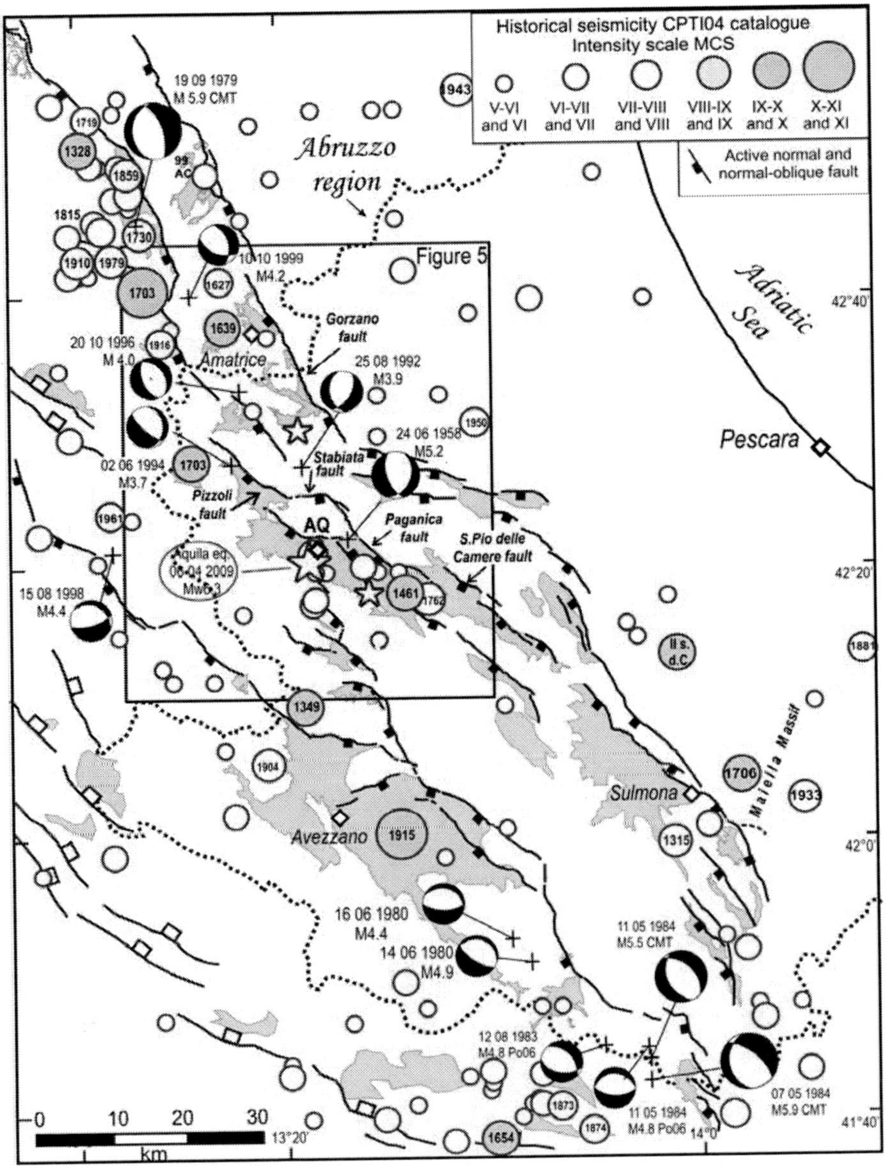

Figure 3. Late Pliocene-Quaternary normal faults and associated syn-sedimentary basins in the Abruzzo region plus epicentres of the moderate to large historical earthquakes (CPTI04 catalogue, Working Group CPTI, 2004) and focal mechanisms of the intra-Apennine instrumental events. The circled numbers indicate the year of the historical earthquake; the instrumental earthquakes are labelled with their date given as day/month/year. Futher explanation can be found in Boncio et al. (2009).

2001; Galli et al., 2008 and references therein). In particular, within and close to the epicentral area of the L'Aquila seismic sequence, two major normal fault systems can be recognised: Mt. Gorzano-Gran Sasso and Pizzoli-Stabiata-Paganica-S.Pio delle Camere (Figure 3).

The Mt. Gorzano SW-dipping normal fault dissects the western limb of a coaxial Early Pliocene anticline, which consists of a Middle-Late Miocene pelagic succession covered by Messinian siliciclastic turbidites. The fault extends for 28 km along-strike with a N140°–150°E mean direction and a prevailing dip-slip kinematics (Boncio et al., 2004b). The along-strike variation of displacement, ranging from some hundred of meters to over 2000 m, constrains the Gorzano fault as an isolated structure, without significant internal segmentation. The displacement and the eastward tilting of the Pleistocene-Holocene continental deposits, filling the Amatrice and Campotosto hanging wall basins, and other geomorphic evidences (triangular facets and well-preserved fault scarps), confirm the present activity of the fault (Blumetti et al., 1993; Galadini and Galli, 2003). Towards SE, the Mt. Gorzano fault intersects the northern segment of the N100°±10°E-striking Gran Sasso normal fault system, the Late Quaternary activity of which is well documented (Giraudi and Frezzotti, 1995; Galadini and Galli, 2000).

The Paganica fault (Aquilano *s.l.* fault in Boncio et al., 2004a; Paganica fault in Pace et al., 2006) is a 13-15 km-long, N130°E-striking fault dipping towards SW and extending, from NW to SE, from the Paganica-Tempera villages to the latitude of San Demetrio nei Vestini (Figure 5). Near Paganica, the fault displaces Quaternary alluvial and colluvial deposits in the hanging wall against Lower-Middle Pleistocene conglomerates in the footwall (Geological Map of Italy, 2005). The fault bounds at NE a rectangular-shaped tectonic depression filled by Late Quaternary alluvial and fan deposits. This depression is limited at SW by an antithetic NE-dipping normal fault, the Bazzano fault, characterised by an evident limestone fault scarp, delimiting Late Quaternary slope debris at the hanging wall. SE of Paganica, the fault is organised in right-stepping segments displacing the Lower-Middle Pleistocene continental succession. NNW of the Tempera village, the Paganica fault joins the NW-SE-striking part of the Mt. Stabiata fault through a ~4 km-long system of previously unrecognised (i.e., before the April 6 earthquake) right-stepping en échelon fault planes dipping towards SW (e.g., Colle Enzano and Mt. Castellano segments). The Mt. Stabiata fault, which also shows evidence of activity in Late Quaternary times (Pizzi and Pieruccini, in press), is an 11 km-long S-dipping fault, with a mean E-W direction, articulated in alternating E-W, NW-SE and ENE-WSW segments. Moving westward, the E-W striking portion of the Mt. Stabiata fault almost reaches the SW-dipping Pizzoli normal fault, which extends for nearly 15 km along-strike and also shows clear evidence of Late Quaternary activity (Galadini and Galli, 2000; Moro et al., 2002).

In total, the previously recognised Paganica fault segment (San Demetrio-Tempera) plus its northern continuation towards Mt. Stabiata extend along-strike for ~19 km. If the NW-SE striking segment of the Mt. Stabiata fault, which is aligned with the Paganica fault and shows evidence of coseismic activation during the April 6 shock (see next section), is also taken into consideration, the total fault length is ~20 km in the NW-SE direction.

Coseismic Ground Effects

Soon after the April 6 L'Aquila earthquake, like many colleagues from other Universities and Institutions, the authors have started a detailed field survey in the earthquake epicentral area and neighbouring region, in order to detect possible evidence of coseismic ground deformation (Boncio et al., submitted). We have identified two major NW-SE trending alignments of coseismic features: one coinciding with the Paganica fault plane and its northward continuation along the NW-SE striking portion of the Stabiata fault, the other located within the hangingwall of the Paganica fault, nearly 1.5 km away from the surface fault trace.

The first alignment, especially well exposed in the village of Paganica and north-northwest-ward of it, was followed with continuity for nearly 10 km; the surveyed coseismic features, with an average N130–140°E direction, consist of linear fissures with centimetric opening (up to ~ 12 cm), en échelon cracks, minor shear planes within the fault damage-zone and rejuvenated fault surfaces with centimetric downthrown (up to ~ 10 cm) to the SW (Figure 4).

The other alignment extends for nearly 4.5 km between the villages of Paganica and San Gregorio within the Late Quaternary hanging wall alluvium of the Paganica fault. It consists of linear open fissures and cracks which form a right-stepping en échelon pattern with the other alignment and which have an average strike of N140°E, locally N110°E. In a few places, coseismic gas emission and small sand volcanoes due to water escape were also observed (Figure 4). In total, the area affected by coseismic ground deformation extends in the NW-SE direction for about 13 km; the collected evidence of coseismic fracturing and faulting point to the Paganica fault and its continuation in the NW-SE striking portion of the Mt. Stabiata fault as the causative structures of the April 6 seismic event.

Figure 4. Field evidence of coseismic effects associated to the April 6 earthquake: a) N120°E-striking rejuvenated fault scarp on footwall Quaternary conglomerates along the Paganica normal fault (opening = 3–4 cm; downthrown = 6 cm); b) N145°E-striking ground fissure between the Paganica and Onna villages (opening = 2 cm); c) sand volcanoes originated after the expulsion of water during the earthquake shaking involving completely saturated fine sediments produced by washing procedures in a gravel quarry along the fissure alignment; d, e, f) NW-SE striking segment of Mt. Stabiata fault plane (400 m north of Piano del Monte locality): the main fault plane, the associated fault-breccia and the Late Pleistocene – Holocene slope debris lying on the fault plane show evidence of coseismic deformation confirmed by the diffuse occurrence of centimetric free faces and coseismic fissures sub-parallel to the main fault plane.

SEISMOTECTONIC CONTEXT

During the last thirty years, the Umbria-Marche-Abruzzo portion of the Apennine extensional belt of Italy, which extends for more than 200 km from Gubbio to Barrea, has undergone a frequent and in some cases strong seismic activity, in relation to a common SW-NE-trending tensional stress axis. This is shown in Figure 1 by the epicentral distribution of recent earthquakes and related focal mechanisms. From north to south, these events are: the Gubbio 1984 (M_w 5.6), Colfiorito 1997-1998 (Sept. 1997 M_w 5.6 and M_w 6.0; Oct. 1997 M_w 5.2 and M_w 5.6; Apr. 1998 M_w 5.1), Norcia 1979 (M_w 5.9), and Barrea 1984 (M_w 5.9) earthquakes (Boncio et al., 2004a and references therein). All the main shocks have normal or normal-oblique kinematics and ruptures previously known as active faults.

Evidence of coseismic surface breaks, not unanimously interpreted as surface faulting, is documented only for the 1997 Colfiorito earthquake sequence (Basili et al., 1998; Cello et al., 1998; Cinti et al., 2000). The maximum depth of the activated seismogenic faults, as inferred from the aftershocks, is 6-7 km for the Gubbio structure, 8-9 km for Colfiorito, 11–12 km for the Norcia structure and, 13–14 km for the Barrea fault in the Sangro valley area.

This instrumental seismicity activated the northern half of the intermediate extensional alignment shown in Figure 1, with the only exception of the Barrea earthquake, which belongs to southern portion of the internal alignment. In the 20th century, the latter was struck by the 1904 Mt. Velino – Magnola (M_w 5.5) and 1915 Avezzano (M_w 7.0) earthquakes (Figure 3).

The portion of the intermediate alignment located between the Norcia 1979 and L'Aquila 2009 earthquakes was activated in January-February 1703 by a SE-propagating multiple-rupture seismic sequence consisting of several shocks with three main events between Norcia, Pizzoli and L'Aquila (January 14 Mw 6.7, January 16 Mw 6.0, February 2 Mw 6.6). The NNW-SSE striking Norcia, Cittareale and Pizzoli faults were almost entirely activated by the large 1703 earthquake sequence. The NW-SE striking Paganica fault is probably responsible for the 1461 (Mw 6.5), 1762 (Mw 5.9) and 1958 (Mw 5.2) earthquakes (Boncio et al. 2004a, b; Pace et al., 2006; Lavecchia et al., this book)

The external extensional alignment has less clear seismogenic behaviour. The Mt. Vettore and Gran Sasso faults have been silent since ancient times, even if large paleoseismic events have been documented (Galadini and Galli, 2003; Galadini et al., 2003). The Amatrice-Campotosto area, in the hanging wall of the Mt. Gorzano fault, was struck by one large historical earthquake in October 1639 (M_w 6.3), probably due to the rupture of the northern portion of the Mt. Gorzano fault. Since when instrumental records have been made available, the area has been affected only by minor earthquakes (M≤4.0) and spread microseismicity (Boncio et al., 2009). In particular, three minor seismic sequences occurred during the last decades (August 1992 maximum M_d 3.9, June 1994 maximum M_d 3.7 and October 1996 maximum M_l 4.0); they can be geometrically linked with the southern portion of the Mt. Gorzano fault, down to 14–16 km (De Luca et al., 2000; Boncio et al., 2004b).

Based on an integrated analysis of seismological and rheological data, Boncio et al. (2009) constrained the brittle-ductile transition and the associated seismicity cut-off at depths of nearly 12–14 km below the more internal L'Aquila, Paganica, Fucino, Barrea areas and of 15–17 km below the Amatrice-Campotosto, Gran Sasso and Sulmona areas. This reconstructed configuration of the seismogenic layer fits rather well the hypocentral depths of the major events of the L'Aquila seismic sequence, which deepens from a maximum depth of ~10 km beneath the L'Aquila area, to maximum depths of ~11 km and ~15 km beneath the Campotosto and Ocre areas, respectively (Figures 2 and 5).

SEISMOTECTONIC INTERPRETATION

The aftershock pattern, the location of the largest magnitude shocks and the field evidence of coseismic ground deformation suggest that two major distinct normal faults have been involved during the April 2009 L'Aquila seismic sequence: the SW-dipping Paganica fault and the WSW-dipping Mt. Gorzano fault (Figures 2 and 5). The aftershock activity seems also to have involved very marginal portions of the Gran Sasso and Pizzoli faults.

The NW-SE Paganica normal fault was entirely activated, from its southern tip, close to the village of San Demetrio nei Vestini, up to the intersection with the E-W striking portion of Mt. Stabiata fault, for a total length of nearly 20 km along strike. Associated coseismic ground effects were recognised for a length of nearly 13 km in the central-northern portion of the seismogenic source.

This picture is consistent with radar interferometry (DInSAR) data (Atzori et al., 2009), which clearly show the subsidence of the Paganica fault hangingwall block. The subsided area is bordered eastward by the surface trace of the SW-dipping Paganica fault and northward by the surface trace of the S-dipping Mt. Stabiata normal fault. Most of the Paganica fault plane was activated and ruptured by the April 6 main shock (M_w 6.3), from the bottom of the local seismogenic layer up to the surface (L'Aquila epicentral area). The April 7 shock (M_w 5.6) completed the rupture of the Paganica fault towards SE (Ocre epicentral area).

Nevertheless, only the main shock produced ground deformation in the form of open fissures and rejuvenated fault planes with associated centimetric displacements corresponding or parallel to the fault trace. This field evidence can be interprerted as the expression of coseismic surface rupture.

The Mt. Gorzano fault was activated on April 9 by two moderate magnitude shocks (M_w 5.4 and 5.3), plus a number of small-to-moderate events in the following days. To date (May 6, 2009) the epicentral distribution covers a large area at the hangingwall of the southern half of the fault (Campotosto epicentral area). No evidence of coseismic faulting has been recognised in the area.

Figure 5, based on the criteria presented in another chapter of this book (Lavecchia et al.), shows the 3D-geometry of the two major individual seismogenic faults activated by the L'Aquila sequence (Paganica and Gorzano faults) in the form of more or less rectangularly-shaped areas (e.g., boxes *sensu* Lavecchia et al., 2002), which represent the plan view projection of the fault planes. The box length (L) is constrained by surface geologic data, whereas the box width (Ws) is more speculative and based on the assumption of the average fault dip angle (A) and fault width (W), which in turn depends on the local depth (D) of the seismogenic layer.

Both the Gorzano and Paganica source parameters had already been computed by Boncio et al (2004a). Those of the Gorzano faults are confirmed by the additional information obtained from the present earthquake (L ~ 28 km, W ~ 19 km; D ~ 15 km; A ~ 50°; Ws ~ 12 km). Those of the Paganica source must be partially updated, especially concerning the continuation of the source to NW, and therefore the total fault length (L), which it is now known to be nearly 20 km. Assuming an average fault plane dip of 50° and a fault detachment at a depth of ~ ~12 km, a width (W) of ~ 15 km and a surface width (Ws) of ~10 km can be computed, which fit quite well the epicentral distribution of the seismic sequence. More in general, both the evaluated sub-surface rupture length (~20 km) and the survied surface rupture length (~13 km) seem to fit well with the empirical scale-law relations for a Mw 6.3 earthquake.

Figure 5 also shows the macroseismic epicentres and the intensity data points ($I_0 \geq$ VII) of the major historical earthquakes, which activated the Gorzano (1693 M_w 6.3) and Paganica (1461 M_w 6.5, 1762 M_w 5.9, 1958 M_w 5.2) seismogenic sources.

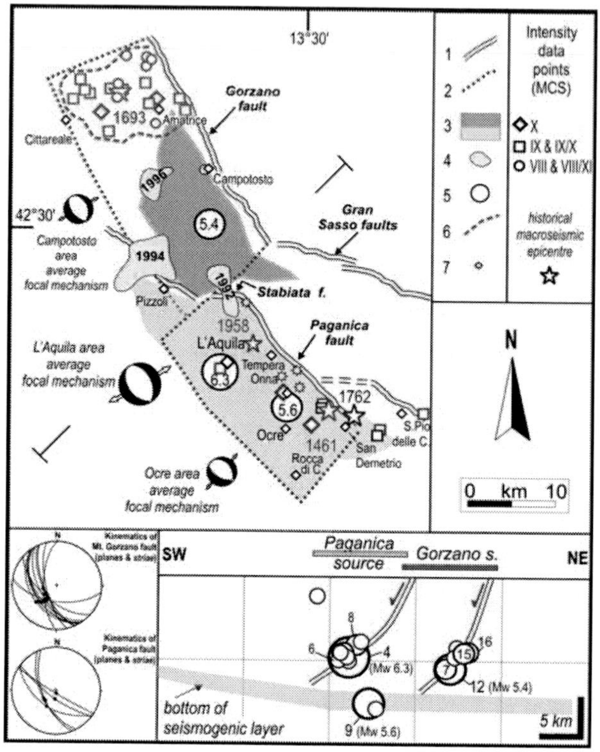

Figure 5. Preliminary seismotectonic interpretation of the L'Aquila 2009 seismic sequence (location of the map in Figure 3). Key: 1) active and possibly seismogenic normal faults in this part of the Abruzzo region (same faults as in Figures 2 and 3); 2) inferred boundary of the seismogenic boxes associated with the Paganica and Mt. Gorzano individual master faults (for the definition of "box" see Lavecchia et al., this book); 3) epicentral areas of the seismic sequences distinguished into a northern portion (green) mainly at the hangingwall of the Mt. Gorzano fault and a southern portion (pink) mainly at the hangingwall of the Paganica fault; 4) epicentral areas of the 1992, 1994 and 1996 minor extensional seismic sequences (from Boncio et al., 2004b); 5) epicentres of the three major events of the L'Aquila seismic sequence (the numbers within the circles give the corresponding M_w); 6) boundary of the seismogenic rupture area of the Amatrice 1693 earthquake (M_w 6.3), from Boncio et al. (2004b); 7) sites of the coseismic deformation shown in Figure 4. The intensity data points refer to the major historical earthquakes (1461 M_w 6.5, 1693 M_w 6.3, 1762 M_w 5.9) (Stucchi et al., 2007). The bottom section indicates the hypocenters of the $Mw \geq 4.0$ events of the L'Aquila sequences compared with the section view geometry of the Paganica and Gorzano fault and the depth of the underlying brittle-ductile transition as defined by Boncio et al. (2009); the hypocenter coordinates were kindly provided by Claudio Chiarabba, INGV.

CONCLUSION

The L'Aquila April 2009 seismic sequence is the largest and most destructive sequence (about 300 casualties) in central Italy since the Avezzano 1915 earthquake. It covers a 45 km-long area elongated in the NNW-SSE direction, released energy of about 1.5E+14J and activated two distinct seismogenic sources: the Paganica and the Mt. Gorzano normal faults. The first, associated with the M_w 6.3 event, was entirely activated for a length of nearly 20 km and a width of about 15 km; the second was activated only in the southern half (mainly south of the town of Campotosto).

The L'Aquila April 2009 seimic sequence, although fitting rather well with previously predicted seismotectonic scenarios, gives us some new insights into the possible patterns of stress interaction among the neighbouring faults of the Apennine seismogenic extensional belt. The previous Italian earthquake history encompasses cases of large events activating a single source, such as the M_w 7.0 1915 Fucino earthquake (Ward and Valensise, 1989), and of multiple events triggering along-strike neighbouring individual sources, such the Colfiorito 1997 (M_w max 6.0) and the Norcia-Aquilano 1703 seismic sequences (M_w max 6.8). On the contrary, the Paganica and Gorzano seismogenic structures, both activated in the course of the L'Aquila seismic sequence, are right-stepping normal faults (with a separation of 7–8 km perpendicular to strike), which belong to two well distinct regional scale active fault alignments (Figure 1). After the activation of the Paganica individual source with the M_w 6.3 event on April 6, the seismogenic process did not propagate along the same intermediate alignment – e.g. along the Pizzoli fault - but jumped north-eastward to the Mt. Gorzano fault of the external alignment (Figures 1 and 5). Consequently, an ongoing process of linkage between the two fault systems under a tensional stress field dominated by WSW-ENE tension may be hypothesized.

REFERENCES

Atzori, S.; Hunstad, I.; Chini, M.; Salvi, S.; Tolomei, C.; Bignami, C.; Stramondo, S.; Trasatti, E.; Antonioli, A.; Boschi, E. *Geophys. Res. Lett.* 2009, *36*, L15305, doi:10.1029/2009GL039293.

Basili, R.; Bosi, C.; Galadini, F.; Galli, P.; Meghraoui, M.; Messina, P., Moro, M.; Sposato, A. J. *Earthq. Engin.* 1998, 2, 291-302.

Blumetti, A. M.; Dramis, F.; Michetti, A. M. *Earth Surf. Process. Landf.* 1993, *18*, 203-223.

Boncio, P.; Lavecchia, G. *J. Geodyn.* 2000, *29*, 233-244.

Boncio, P.; Lavecchia, G.; Pace, B. *J. Seismol.* 2004a, *8/3*, 407-425.

Boncio, P.; Lavecchia, G.; Milana, G.; Rozzi, B. *Ann. Geophys.* 2004b, *47*, 1723–1742.

Boncio, P.; Pizzi, A.; Brozzetti, F.; Pomposo, G.; Lavecchia, G.; Di Naccio, D.; Ferrarini, F. *Geophys. Res. Lett.,*submitted.

Boncio, P.; Tinari, D.P.; Lavecchia, G.; Visini, F.; Milana, G. *Boll. Soc. Geol. It. (Ital. J.Geosci.)* 2009, *128*, 367-380.

Brozzetti, F.; Boncio, P.; Lavecchia, G.; Pace, B. *Tectonophysics.* 2009, *463*, 31-46.

Carey-Gailhardis, E.; Mercier, J. L. *Earth Planet. Sci. Lett.* 1987, *82*, 165-179.

Cello, G.; Deiana, G.; Mangano, P.; Mazzoli, S.; Tondi, E.; Ferreli, L.; Maschio, L.; Michetti, A. M.; Serva, L.; Vittori, E. *J. Earthqu. Eng.* 1998, *2*, 303-324.

Cinti, F. R.; Cucci, L.; Marra, F.; Montone, P. *J. Seismol.* 2000, *4*, 333–343.

De Luca, G.; Scarpa, R.; Filippi, L.; Gorini, A.; Marcucci, S.; Marsan, P.; Milana, G; Zambonelli, E. *J. Seismol.* 2000, *4*, 1–21.

Galadini, F.; Galli, P. *Nat. Hazards.* 2000, *22*, 225-270.

Galadini, F.; Galli, P. *Ann. Geophys.* 2003, *46(5)*, 815-836.

Galadini, F.; Galli, P.; Moro, M. *Ann. Geophys.* 2003, *5*, 793-813.

Galli, P.; Galadini, F.; Pantosti, D. *Earth-Sci. Rev* 2008, *88*, 89-117.

Geological Map of Italy, *Sheet 359 L'Aquila, scale 1:50.000* 2005, APAT, Servizio Geologico Italiano e Regione Abruzzo.

Giraudi, C.; Frezzotti, M. *Quat. Int.* 1995, *25*, 81-93.

Lavecchia, G.; Boncio, P.; Brozzetti, F.; Stucchi, M.; Leschiutta, I. *Mem. Soc. Geol. It.* 2002, *1*, 881-890.

Moro, M.; Galadini, F.; Galli, P.; Bosi, V.; Giaccio, B.; Messina, P.; Sposato, A. *Il Quaternario (It. J. .Quatern. Sci.)*, 2002, *15(2)*, 259-270.

Pace, B.; Peruzza, L.; Lavecchia, G.; Boncio, P. *Bull. Seism. Soc. Am.* 2006, *96*, 107-132.

Pizzi, A.; Galadini, F. *Tectonophysics.* 2009, *476*, 304–319.

Pizzi, A.; Calamita, F.; Coltorti, M.; Pieruccini, P. *Boll. Soc. Geol. It.* 2002, *Spec. Vol. 1*, 923-929.

Pizzi, A.; Pieruccini, P. *Note illustrative della Carta Geologica d'Italia*, in press, ISPRA Servizio Geologico d'Italia.

Stucchi, M.; Camassi, R.; Rovida, A.; Locati, M.; Ercolani, E.; Meletti, C.; Migliavacca, P.; Bernardini, F.; Azzaro, R. *Quaderni di Geofisica, INGV,* 2007, *49*, 38 pp., available at: http://emidius.mi.ingv.it/DBMI04/.

Valensise, G.; Pantosti, D. (eds.) *Ann. Geophys*. 2001, *44 (Suppl.)*, 180 pp. with CD-ROM.

Vezzani, L.; Festa, A; Ghisetti, F.. *Geological-structural map of central-southern Apennines (Italy),* 2009, Sheet 1, scale 1: 250.000.

Ward, S. N.; Valensise, G.R. *Bull. Seism. Soc. Am.* 1989, *79*, 690-710.

Working Group CPTI, *Catalogo Parametrico dei Terremoti Italiani* 2004, available at http://emidius.mi.ingv.it/CPTI04/.

In: Recent Progress on Earthquake Geology
Editor: Pierpaolo Guarnieri, pp. 19-48
ISBN: 978-1-60876-147-0

Chapter 2

EARTHQUAKE GEOLOGY OF THE KACHCHH REGION, NORTHWEST INDIA

C. P. Rajendran and Kusala Rajendran
Centre for Earth Sciences, Indian Institute of Science,
Bangalore 560012, India.

ABSTRACT

Geological methods are extremely useful in characterizing earthquake source zones and estimating recurrence intervals. Recent advances in paleoseismological research have led to various spatial and temporal models that are fundamental to the understanding of earthquake recurrence. These models, however, are region-specific and most regions where studies have taken place have not generated large earthquakes in the recent times. Kachchh region of northwest India offers a unique setting in this regard because of the occurrence of two large earthquakes, in 1819 and 2001. Although both these events occurred by the reactivation of an ancient rift, they are different in terms of surface expressions of deformation and recurrence intervals. The 1819 earthquake occurred on the northern boundary fault of the Kachchh rift, and it produced significant elevation changes along the ~90 km-long stretch of the fault. The 2001 Bhuj earthquake occurred on a smaller fault within the rift and was associated with a deeper source. The 2001 earthquake generated secondary features, such as sandblows, lateral spreads and secondary faults, which have served as proxies of past events. Paleoseismic studies in these regions suggest varying recurrence intervals, ~1000 and ~4000 years, respectively for the 1819 and 2001 sources. Large earthquakes in this region appear to be associated with multiple sources and variable recurrence intervals compared to other seismogenic paleo-rifts. We attribute the higher frequency of earthquakes associated with this rift to the proximity to an active plate boundary, the relatively thinner Indian lithosphere and the mid-lower crustal rift pillows. These characteristics make the Kachchh different from New Madrid and Charleston, all of which are failed rift systems and/or extended crust associated with large earthquakes. This paper presents an evaluation of the paleoseismologic investigations in the Kachchh region, leading to the estimation of timing of past events.

Keywords: paleoseismology, earthquake recurrence, liquefaction feature, Kachchh rift.

INTRODUCTION

Investigations in fault zones around the world suggest that the geological methods provide useful constraints to understand the long-term behavior of seismogenic structures, and it marks an important deviation from the traditional approaches in seismic hazard assessment. Even in countries with longer historical records, useful information on the seismic activities may not exist because the recurrence intervals of moderate/large earthquakes are often longer than the historic records. Palaeoseismologic and geologic data provide useful inputs for recurrence models in different tectonic environments. Studies in some of these fault zones have suggested recurrence of characteristic slip, separated by a few thousands of years [Schwartz and Coppersmith, 1984], while some others such as the San Andreas fault in the western United States suggest recurrence intervals of the order of a few hundreds of years [Sieh, 1978; Grant and Sieh, 1994]. Studies in less active cratonic regions such as Australia and peninsular shield region of India indicate earthquakes separated by intervals of many thousands of years [Rajendran et al., 1996; Crone et al., 1997]. Studies have also helped in identifying paleoearthquakes in regions such as the Rhine valley, which has not been significantly active in the historic or recent times [Meghraoui et al., 2001]. The science of earthquake geology is now being increasingly used to assess the history of seismic source zones and to predict the size and deformational characteristics of the future events [McCalpin, 1996; Yeats et al., 1997; Noller et al., 1999].

While active fault systems such as the San Andreas fault in the western United States may be evaluated based on more recent activity, many seismic source zones around the world are not associated with any perceptible recent seismicity. The seismic source zones associated with some of the paleorifts

(Kachchh rift, India) and passive margins (Charleston, South Carolina, USA) may be cited as examples. There are also many other examples of moderate earthquakes that have occurred in regions of low background seismicity where geological investigations have revealed elapsed times of the order of several thousands of years, giving a general perception of aseismicity [Rajendran et al., 1996; Crone et al., 1997]. There is also the example of the New Madrid Seismic Zone (NMSZ) in the SE United States, which seems to be generating large earthquakes over intervals of a few hundred years, although this rate remains debated [Newman et al., 1999; Tuttle and Schweig, 1991, Tuttle et al., 2002]. Occurrence of two large earthquakes (Mw 7.6) during a short span of 180 years (1819 and 2001) gives an exceptional status to the Kachchh seismic zone, Gujarat, (figure 1) considered to be a part of the stable peninsular shield of India [Johnston, 1996]. The 1819 earthquake is known for the spectacular ground uplift known as the "*Allah Bund*" that found mention even in the early geology books [Lyell, 1857]. This elevated landform and other coseismic features have remained largely unexplored for a long time; proximity to a sensitive international border, remoteness and inclement weather conditions being the main deterrents. However, due to its isolated location, semi-desert conditions and low-level human activity, coseismic features are expected to be reasonably well-preserved. Our studies on the 1819 earthquake were based primarily on the historic data, landform changes and seismically induced near and far field liquefaction features [Rajendran and Rajendran, 2001].

The 2001 earthquake, the largest paleorift-related event to have occurred since the advent of modern seismology, is well studied, both from the seismological and geological perspectives [Mandal et al., 2004; Bodin and Horton, 2004; Rajendran et al., 2008]. Unlike the 1819 event, the 2001 earthquake did not generate any primary surface rupture but it caused liquefaction and land failure, giving an opportunity to validate inferences on the historical 1819 earthquake. For the 2001 event we used liquefaction features as well as secondary faults to identify the past events and perhaps, this study provided the first example in using the secondary faults to infer previous faulting events [Rajendran et al., 2008]. Our exploration in this region has uncovered seismically induced paleoliquefaction features, which seem to be spatially and temporally independent of these two known sources, based on which we proposed existence of multiple sources with varying recurrence and styles of deformation [Rajendran et al., 2008]. In this paper we demonstrate the usefulness of seismically induced liquefaction features and secondary faults to reconstruct the spatial and temporal characteristics of past earthquakes.

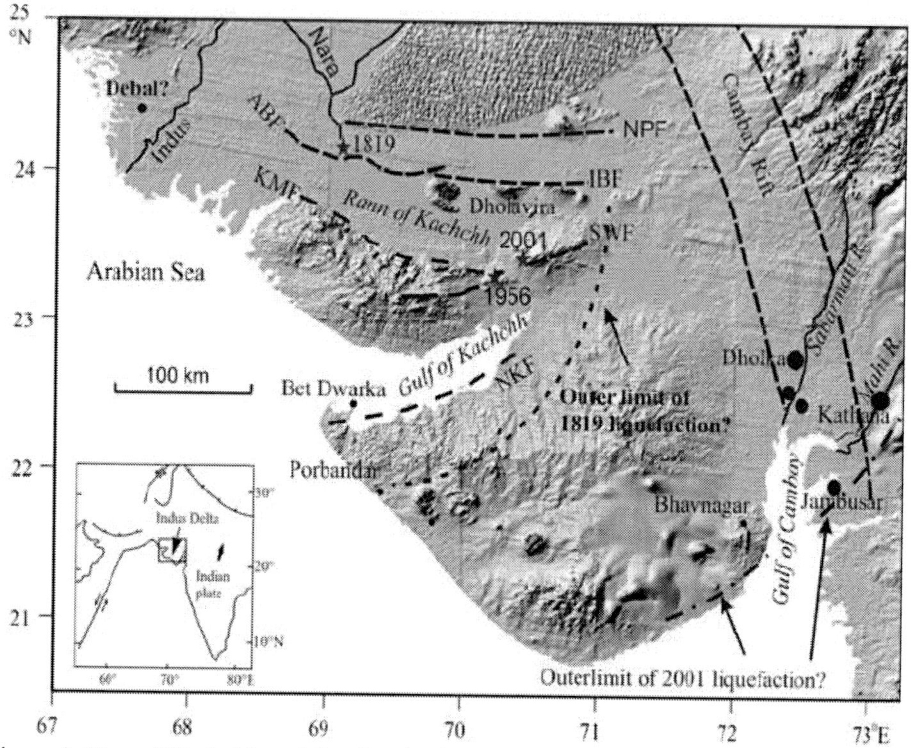

Figure 1. Map of Kachchh and Cambay basins showing the locations of majorearthquakes and faults. The dashed lines indicate the farthest liquefaction limit during the 2001 Bhuj earthquake. NPK-Nagar Parkar fault; ABF-Allah Bund fault; IBF- Island belt fault; KMF-Kachchh mainland fault; KTF-Katrol fault; NKF-North Kathiawar fault; L: Lothal; D: Dholka; M: Mothibaru; J: Jambusar; K: Kathana; MR- Mahi River; SR- Sabarmati River. NPF and NKF form the boundaries of the Kachchh rift. Inset: Map showing the study area; arrow shows the direction of the plate movement.

The 1819 Allah Bund Earthquake

The earthquake that occurred on 16 June 1819 generated a ~ 90-km-long scarp (*Allah Bund*), with a maximum coseismic uplift of about 4.3 m and width of about 15 km, most of which are still preserved (figure 2).

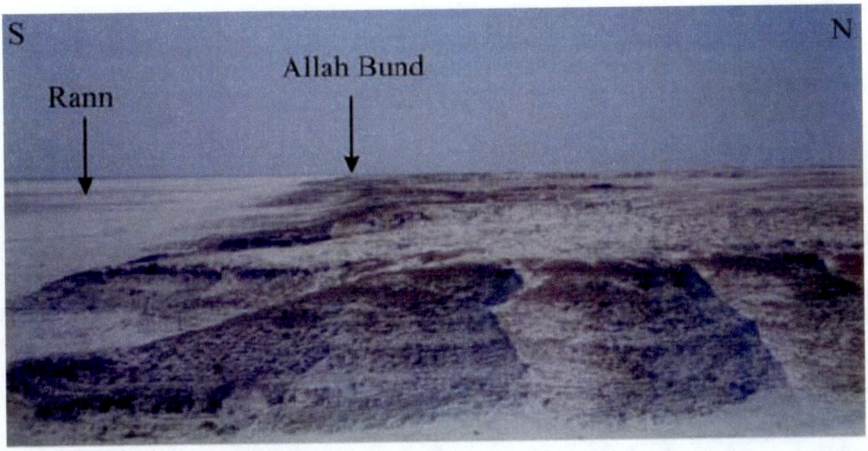

Figure 2. View of the Allah Bund, an uplifted tract of land in the Rann of Kachchh associated with the 1819 earthquake.

Coseismic changes associated with this earthquake are reported to have completely destroyed several fortified townships and permanently altered the local drainage system [Oldham, 1926]. The rich documentation of the coseismic effects, reliable historic records, existence of a four-five thousand year old civilization, well-preserved expressions of ground deformation and high liquefaction potential make this a unique site for paleoseismologic studies. The effects of the 1819 earthquake, including the farthest liquefaction at a distance of 250 km in the town of Porbunder have been quite well documented [Oldham, 1926]. At the town of Vigakot the Nara River, a distributary of the Indus River was dammed due to changes in land level. Most remarkably, the earthquake-related offset uplifted part of the Rann for almost a distance of 80-90 km in the east-west direction. The scarp now appears to have been declined much due to erosion [see Rajendran and Rajendran, 2001, for a discussion on Allah Bund].

NEAR SOURCE STUDIES

The 1819 earthquake generally referred to in the literature as the Allah Bund earthquake led to large-scale liquefaction in the source zone [Mac Murdo, 1824]. Liquefaction combined with downfaulting is reported to have caused extensive flooding that the people who took refuge in the fort at Sindri, located 8 km south of Allah Bund had to be rescued in boats, the following day. Rajendran and

Rajendran [2001] give a detailed account of the coseismic effects of the 2001 earthquake. A fort at Vigakot, located within a distance of 10 km from the inferred epicenter of the 1819 earthquake was also destroyed. The ruins of the Vigakot Fort are still strewn around the region (figure 3a). This is where we started our first search for the evidence of previous destruction.

Figure 3a. Remains of an old fort destroyed in the 1819 earthquake at Vigakot.

Trenching excavations in this region exposed remains of another brick foundation at a depth of >1.5 m, which we consider as the evidence for an older settlement destroyed by an earthquake. The older sandblows have cut through this older settlement horizon, as evident from the bricks and charcoal obtained from the layer immediately below the sandblow. The trenches exposed paleo-sandblows belonging to two generations- the 1819 and its predecessor of A.D. 885-1025, a result obtained from the radiometric carbon dating of the charcoal obtained from the sites. Below the pre-1819 sandblow crater we observed faint signatures of a much older liquefaction feature, for which we could not obtain any age data (figure 3b).

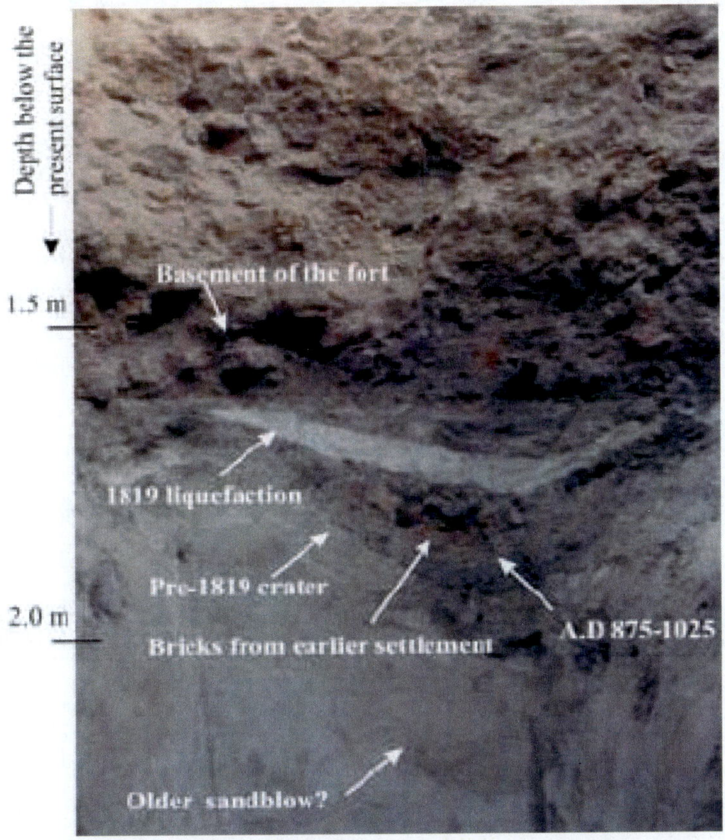

Figure 3b. Photograph of the multiple phases of liquefaction from the western wall of a 2-m-deep trench at Vigakot. The younger liquefaction event composed of clean white sand is intruded into an older sandblow crater, which contains bricks from an older settlement, along with bones, pottery, and charcoal. Basement of the fort destroyed in 1819 is exposed 1.5 m below the surface. Faint outline of an earlier sandblow is exposed below this crater, but this could not be dated.

A nearby trench at Vigakot exposed vertical sand dikes that have cut through the pre-1819 settlement level (see figure 4a). Thus, the trenches in Vigakot region exposed two generations of sandblows, belonging to 1819 and older one about 1000 yr BP and a third one for which age constraints are yet to be made. The 2001 earthquake generated numerous sandblows, the surface expression of a typical small-scale sandblow formed in the Rann is shown in figure 4b. At the time of publication of the results from the 1819 source zone, the 2001 Bhuj earthquake occurred, and the whole region of Rann including part of Allah Bund was impacted by liquefaction. Large sandblows generated by the 2001 earthquake in the close proximity of Vigakot (inferred epicenter of the 1819 earthquake), exposed two older generations of sand blows and related vents and cross cutting features. Three generations of sand venting (2001, 1819 and older event ~ 1000 yr BP) could be documented in the same trench, with the maximum sandblow thickness for the 1819 event [see Rajendran and Rajendran, 2002]. While the 1819 and the penultimate earthquake appeared closer to this site, the source of the older earthquake remained elusive. Identification of sources in the Kachchh region is a difficult problem due to the high liquefaction potential and the fact that the large earthquakes could affect distant locations [Rajendran et al., 2001]. A rare historic document discussing the effects of the AD 893 Debal earthquake helped resolve this issue.

Figure 4a. View of a sandblow formed in the Rann during the 2001 Bhuj earthquake.

Figure 4b. A sand dike, possibly from the 1819 earthquake cuts through an occupation level marked by bricks and pottery.

Careful interpretation of some historic documents led to the conclusion that the ~1000 year old earthquake observed in the trenches close to Vigakot was probably the AD 893 earthquake that occurred close to Debal, an ancient port in the Indus Delta [Rajendran and Rajendran, 2003; figure 1]. Although historic accounts suggested the occurrence of this earthquake in the Indus Delta, its location remains uncertain, also considering the fact that large areas of high liquefaction susceptibility are affected by each event. For example, the 2001 earthquake generated sandblows resulting in craters measuring 5 m across, forming 10-15 cm thick sand layers, at locations close to Vigakot, about 80 km away from the source. Similar features have also been reported from Hyderabad, in Pakistan. However, the trenches exposing the three events clearly showed the difference in thickness of the sand layers originating from the near and far events. For example, excavation of the 2001 craters suggested that 30-35 cm-thick sand deposits from the 1819 earthquake are still preserved in some sections, but the maximum thickness sand deposited by the 2001 sandblows at the same location is about 10 cm [Rajendran et al., 2001].

In addition to obtaining age data on the penultimate event and exposing several potential sites for excavation, studies on the 1819 earthquake provided details on the style of deformation. The coseismic uplift evidenced by Allah Bund has generally been considered as an expression of the fault, the block south of the bund moving down with respect to the northern block [Bilham, 1999]. We also started with the presumption that this feature could be the expression of the fault and our early investigations involved trenching excavations across this structure.

More than 10 of the trenches (2-3 m deep) did not suggest any evidence of visible fault offset. We reported morphological features such as progressive migration of streams and formation of a lake at the down dip side of the upthrust block, which we interpreted as indications of large-scale folding [Rajendran and Rajendran, 2001]. Thus, we suggested that the south verging 3-4-m-high Allah Bund could be the eroded face of a fold, formed by the 1819 earthquake.

Analyses of focal mechanisms of the aftershocks of the 2001 earthquake suggest predominantly compressive stress regime, with the maximum principal stress oriented in the N-S direction [Antoliok and Dreger, 2003; Mandal, 2008]. The 2001 earthquake and continuing aftershocks have occurred on a younger imbricate thrust within the rift, whereas the 1819 earthquake seems to be related to the northern boundary fault of the Kachchh rift. Reactivation of the fault system in the post-collisional stress regime may be the controlling mechanism for these earthquakes. The inferred spatial pattern of paleoliquefaction features suggests evidence for older earthquakes and previously unknown sources.

Distant Liquefactions and Older Events

The sediment-rich environment in the Rann combined with the shallow water table is regarded as highly favorable for liquefaction. The Bhuj earthquake generated liquefaction features in locations as far as 300 km from the epicenter of the earthquake; there were reports of spouting of water near Jambusar, Broach District, located about 300 km from the epicenter. Liquefaction and venting of sand were reported also from Hyderabad (in Pakistan), located about 300 km northwest of the epicenter. Other sites of liquefaction features are observed also near Dholka, Mothibaru and Lothal, about 200 km, southeast of the 2001 epicenter. These are some of the sites excavated by us in our search for earlier episodes of liquefactions. We conducted independent trenching work in some sites in Cambay basin (e.g. Kathana and Bhavnagar) and at a site near Dwarka near North Kathiawar Fault (NKF) near Dwarka, where we discovered indications of previous large earthquakes (see figure 1 for locations).

Observations elsewhere suggest that the older sand vents are prone to be reactivated by later events, possibly due to favorable site conditions as well as availability of preexisting fissures [Sims and Garvin, 1995]. This was the rationale in our search for older features associated with the freshly formed features. Due to identical geological and hydrological conditions in most low-lying parts of Kachchh, we expect that the characteristics of liquefaction (source–distance relation and size of sand blows) would be more or less comparable for similar-size events. We also used this reasoning to calibrate the size of past events and speculate on their probable locations. Two distant sites investigated by us lead to identification of older events independent of the 1819 and 2001 sources and possibly associated with a different temporal behavior. In the following sections, we discuss these aspects.

Evidence for a Paleo-Event at ~3000 Yr BP: Dholka and Kathana (Cambay Basin)

One of the farthest liquefaction features that we noted during the post-2001 surveys was near the village of Dholka, near Ahmedabad, about 250 km from the source of the earthquake [Rajendran et al., 2002]. The amount of sand and water ejected was much smaller compared to the liquefaction features closer to the source. At the time of our studies in this area (early April 2001) the water had dried up, leaving white patches of salt left by the ejected water. The top soil in the plough field that we investigated composed mainly of 0.5 m-thick black cotton soil, underlain by fine white sand, possibly a paleo river channel. The 2001 ejection had brought out this fine sand, mixed with black clay, forming a thin deposit (maximum thickness: 5 cm). The feeder dikes could be traced to the source bed. The trench sections also revealed out-of-sequence deposits of white sand, derived from the source bed below (figure 5), which we interpret as an earlier liquefaction event. This earlier sand blow deposit, which we interpret as the penultimate event affecting this source, is only a few centimeters thick (maximum of 5 cm) and is disturbed and discontinuous. Luminescence dating of the source sand below and the emplaced sand yielded dates of 5138□514 yr BP and 2945□295 yr BP, based on which we infer that the previous event that affected this region must have occurred about 3000 years ago.

Encouraged by the evidence for paleoliquefaction features in the vicinity of the Sabarmati River, we excavated more trenches to the east of this site, on the western banks of the Mahi River (see figure 1 for location). Five trenches were excavated near Kathana, three of which were 2-m-deep. The trenches exposed a series of sand-filled conduits (5-8 cm wide), extending from the bottom of the trench, cutting through the host sediment consisting of thick black clay (1-1.5 m thick). Some of them are linked to sand-filled craters (width of 30-50 cm) toward the upper level. The fine white sand within the craters are 30-50 cm thick; contains carbonaceous matter with clasts of the black clay and also shows flow structures, suggesting forceful ejection. The carbonaceous sediments collected from within one of the craters gave an age of 2353-2730 cal yr BP, in the same range as the date from Dholka. With thicker layers of vented sand and well-developed sand-filled craters, Kathana appears to be closer to the source than Dholka (both located on river beds of comparable geological and hydrological conditions). Thus, we interpret the source to be unrelated to the 2001 Bhuj source and suggest that source may well be within the Cambay basin [Rajendran et al., 2008].

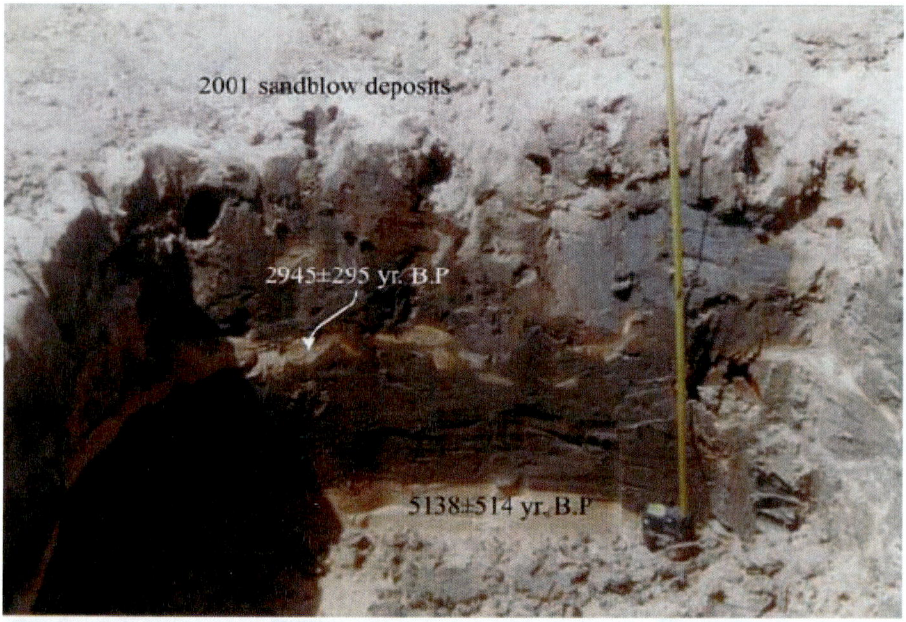

Figure 5. View of a shallow trench in a plough field near Dholka exposing a pre-2001 sand emplacement. The emplaced sand layer appears discontinuous and broken. The 2001 sand blows deposited about 5 cm thick layer of whitish, salty sand which forms the surface layer in this field.

An independent study in the Mahi River valley suggests occurrence of an ancient earthquake between 3369-3729 cal yr BP and 2840-3265 cal yr BP [Maurya et al., 1998]; the younger age corresponds to the maximum age obtained from Kathana sand blow. The sand dikes observed in the sections near Bhavnagar (figure 1) cuts through the late Harappan occupation level (3350-3850 yr BP) providing additional constraints to a liquefaction event that post-dates the occupation level. Interestingly, a historical event of moderate size has been reported from this region (AD 1705) by Iyengar et al. [1999], adding credence to the argument for an active seismogenic source in this region.

EVIDENCE FOR A PALEO-EVENT AT ~2000 YR BP: BET DWARKA (GULF OF KACHCHH)

Bet Dwarka is a narrow strip of island in the Gulf of Kachchh, located 5 km north of the mainland (figure 1); where the 2001 earthquake produced minor damage. A locale of ancient habitation, Bet Dwarka was a port town since Harappan times (~3000 yr BP). A major fault is located close to this area (figure 1), but no significant seismic activity has been reported from here. Archaeological excavations in the region subsequent to the 2001 earthquake revealed multiple levels of settlements, which have been dated using artifacts of pottery shards belonging to different vintages (~ 3800 yr BP onwards), as reported by Vora et al., [2002]. The possibility of finding event horizons with better chronological constrains is one factor that makes this site useful.

In one of the trenches excavated in this region we identified a feature that was interpreted as a sandblow formed about 2000 yr BP [Rajendran et al., 2003]. The wedge-shaped layer of laminated silty sand with pottery, shells, bones and charcoal with a maximum thickness of about 20 cm, was found at a depth of 2.75 to 3 m from the surface (figure 6a,b). This discordant layer of sand whose source vent could not be traced, featured flow structures and rip-up clasts suggestive of forceful ejection of water. This layer occurred at the top of the lowermost horizon, suggesting that humans had occupied this region at the time of the event [Rajendran et al., 2003a]. Charcoal samples from within and immediately outside the emplaced sand layer, provided a radiocarbon age of 1980 ± 40 yr BP (cal. BC 50–AD 100; figure 5), whereas the sample obtained from the sand body yielded a slightly younger age of 1859 ± 60 yr BP (cal. AD 25–328). Artifacts and pottery collected from the layer immediately below the emplaced sandy layer suggest an age of BC 200–AD 200 (~ 2200 yr BP), which we consider as the maximum age of the liquefaction episode.

Figure 6. a. Composite photograph showing the ancient sand blow on the northern wall, formed over the oldest occupation level (level 11) exposed at Bet Dwarka. The section shown here is at a depth of 2 m from the surface. (Inset) Part of the sand blow magnified to show rip-up clasts; bar scale is 5 cm; b. Sketch of the northern trench wall showing level 11 and the sandblow deposit.

The region does not show any evidence of liquefaction in the younger strata nor any younger dikes that crosscut the ~ 2000-year-old sand blow, suggesting that events of comparable dimension may not have affected this area during the subsequent periods. The other shallower trenches (1.5–2 m) that we excavated in this area also did not expose any younger liquefaction features. Rajendran et al. [2003b] suggested two possible scenarios to attribute a possible source to the 2000-year-old feature. One, it might be an outlier of a distant seismic source, and two, it might be related to a nearby source. The first scenario immediately raises the question if the feature was associated with the 1819 or 2001 sources. The absence of any contemporaneous features (~2000 yr BP) in both these source regions excludes that possibility. Further, the emplaced layer at Bet Dwarka has not been cross-cut or disturbed by later events, an indication that this site is outside the probable limits of large-scale liquefaction due to earthquakes originating from the two known sources (the 1819 and the 2001 events). The Bhuj earthquake whose epicenter is about 200 km away from Bet Dwarka did not generate any noticeable liquefaction even in the low-lying areas of Dwarka, bearing testimony to the above argument. These observations led Rajendran et al (2003b) to conclude that the ~ 2000 year old event identified from the trenches in Bet Dwarka may have originated from an independent source, which needs to be constrained from future studies. Studies in the source zone of 2001 earthquake and other potential locations within the Cambay basin have led to identification of a few past earthquakes, which are discussed in the following sections. However, none of them can be bracketed with the 2000-year-old event at Bet Dwarka and we suggest that this event may have originated from a hitherto unknown source.

Investigations in the Near Source Region of the 2001 Earthquake

The 2001 earthquake although named after Bhuj, the largest town that was affected by the earthquake, is located about 70 km NE, in the village of Bachau [Kayal et al., 2002; Antoliok and Dregor, 2003; Bodin and Horton, 2004; Mandal et al., 2004; Singh et al., 2004; figure 1]. Although the location is spatially close to the north-dipping, Kachchh Mainland Fault, the earthquake was related to a steep south-dipping blind fault well-within the rift. The surface projection of the fault plane coincides with a location 20-25 km north of the Kachchh mainland fault [Mandal et al., 2004; Gahalaut and Burgmann, 2004]. The aftershocks are confined to a relatively small area (60 X 40 sq. km) and the focal depths range from 10 to 45 km, defining a 45-50 degree south-dipping westerly striking plane [Bodin and Horton, 2004; Mandal et al., 2004]. Despite its magnitude (Mw 7.7), the 2001 earthquake did not generate any surface scarp, unlike the 1819 earthquake. The surface deformation essentially consisted of mild flexures, folds, extensional cracks, and tear faults, in addition widespread liquefaction features that included sandblows and lateral spreads [Rajendran et al., 2001; Tuttle et al., 2002; Schweig et al., 2003]. During the post-seismic surveys we investigated several of these features, as far as 100 km from the source of the earthquake, and some features were selected for detailed studies in our search for paleoearthquakes. To our knowledge, the study by Rajendran et al. [2008] is the first of its kind where coseismic features, including secondary faults of a recent continental earthquake have been used as proxies of similar events in the past.

Lateral Spread at Budharmora

The 2001 earthquake led to lateral spreads at many locations caused by liquefaction of substrata on very gentle slopes and subsequent down-slope movement towards the toe. The conditions in which such sand dikes form and their usefulness in paleoseismological studies have been discussed widely [Obermeier, 1994; McCalpin, 1996]. Ground deformation observed in an agricultural field at Budharmora located close to the epicenter consisted of 1-m wide extensional cracks and back-rotated soil blocks, associated with extrusion of sand and shortening at the toe (figure 7). We expected to see evidence of similar past events and thus excavated several trenches across the extensional cracks and ridges [see Rajendran et al., 2008 for details]. The trenches exposed three generations of feeder dikes, the oldest of which has broken through an erosional contact, about 2 m below the present ground level, but does not make to the top layers (figure 8). The second set of feeder dikes has cut through the same erosional surface and terminates at the top of a cultural horizon marked by pottery and bricks. The third and the youngest of the sand dikes originate from this occupational level. The maximum age of the oldest event (ca 10.1 and 13.8 Ka) was obtained by dating the top layer that has been extruded. We do not have a site-specific age for the horizon disturbed by the second generation of dikes, but archeologists have dated it as ranging between 3850-4700 yr BP [Joshi and Bisht, 1994].

Another 3-m-deep trench located about 50 m east of the site discussed above also exposed an earlier generation of sand dikes, terminating at the base of an occupational level. The settlement horizon occurs above a calcrete basement, below which numerous sand dikes abruptly terminate. This settlement horizon is inferred to be of the same age (3850-4700 yr BP) as observed in the earlier trench. The calcrete was dated at 6750-7180 yr BP, which could be the maximum age of the earthquake, but based on the younger age of the settlement that appear to have been disturbed, the maximum age of the earthquake could be considered closer to 4700 yr BP. Thus, the trenches at Budharmora suggested the possibility for two previous episodes of lateral spreads, the older event dated about 10-13K years and the younger and the penultimate event dating to about 4700 yr BP.

Figure 7. Areal view of the lateral spread at Budharmora. Note the extensional cracks that have displaced the drainage pipe.

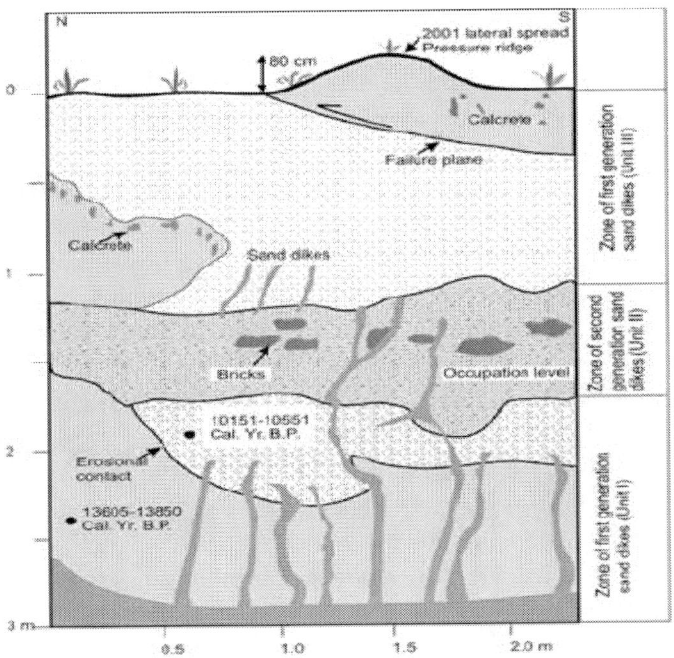

Figure 8. Log of a trench excavated across the lateral spread at Budharmora, showing three generations of sand dikes. The second event has disturbed an occupation level.

Strike-Slip Fault at Manfara

The 2001 earthquake generated an 11-km-long, northwest-striking fault referred to as the Manfara fault [Wesnounsky et al., 2002], near the village of Manfara, which they describe as a 'tear fault' on the hanging wall. We mapped about 3 km length of this fault in an agricultural field across the Kharoi-Chobari road (figure 9). The N 10°E to N 20°W oriented fault displays right-lateral strike-slip, with a maximum lateral displacement of 32 cm [Rajendran et al., 2001]. Rupture is mostly characterized by a series of right-stepping en échelon fractures showing right lateral movement, with a maximum horizontal offset of ~16 cm. An east-west flowing river (*Chang Nadi*), north of the rupture zone takes a northwesterly swing near the Manfara Village. The rupture zone that we mapped follows the same trend as the river for about 1 km, possibly suggesting a long-standing structural control. Study of the aftershocks of the 2001 earthquake by Mandal et al. [2004] suggested at least half a dozen focal mechanism solutions showing strike-slip movement along almost vertical planes, near Manfara. With their focal depths >10 km, these earthquakes appear to be associated with deeper sources and the tear faults may be considered as manifestations of movement at depth. If this rupture was indeed the expressions of a preexisting structure, then we expected to see evidence for previous movements, and this was our rationale for the trenching investigations across this secondary fault. Observations in three trenches (2.5-3 m deep) that we excavated in this region are summarized below.

The shallow stratigraphy consists of indurated medium grained sand, generally uniform and devoid of structures, overlain by brown sandy soil possibly defining an erosional contact between them. The ~1.5-m-thick, brown sandy soil, with patches of the underlying sandstone appears to be part of the plough zone. Walls of trench-1 exposed vertical fissures, as well as a network of sand dikes, that abruptly truncate at the top of the lower sandstone unit (figure 10a). Their upward propagation and the sand emplacement are suggestive of earthquake-induced origin, possibly from an older earthquake in the vicinity. These fissure fills differ from the host rock as they contain very fine-grained pure brownish fine sand with the dikes extending downwards to deeper levels into the source sand reservoir. They are considered to represent the initial phase of development of craters, but too weak to break through the overlying material. Such sand-filled fissures are termed as 'incipient craters' and an earthquake origin is generally ascribed to them [Obermeier, 1996]. They are considered to represent the initial phase of development of craters, but too weak to break the overlying material. The 2001 earthquake, however, did not create any sand injection at this site, despite its large magnitude and proximity to the source [Rajendran et al., 2001].

Some "dry holes", measuring a few centimeters across had appeared in the field, which we interpreted as manifestations of sand injection that failed to reach the surface [Rajendran et al., 2001]. Apparently the ground conditions, especially the depth to water table, appear to have changed since the last event. Some previous studies in the region indicate that wetter environment prevailed until about 4000-5000 years ago [Gaur and Vora, 1999]. Additional evidence for seismically induced features comes from the adjoining trench-2, excavated 6 m south of trench-1. At about 2 m depth, this trench exposed a bowl-shaped crater filled with fining upward sequence of brownish sand and a vertical feeder dike (figure 10b). The ~15-cm-wide crater contains a few small clasts at the margins, suggestive of forceful intrusion of water laden sediments.

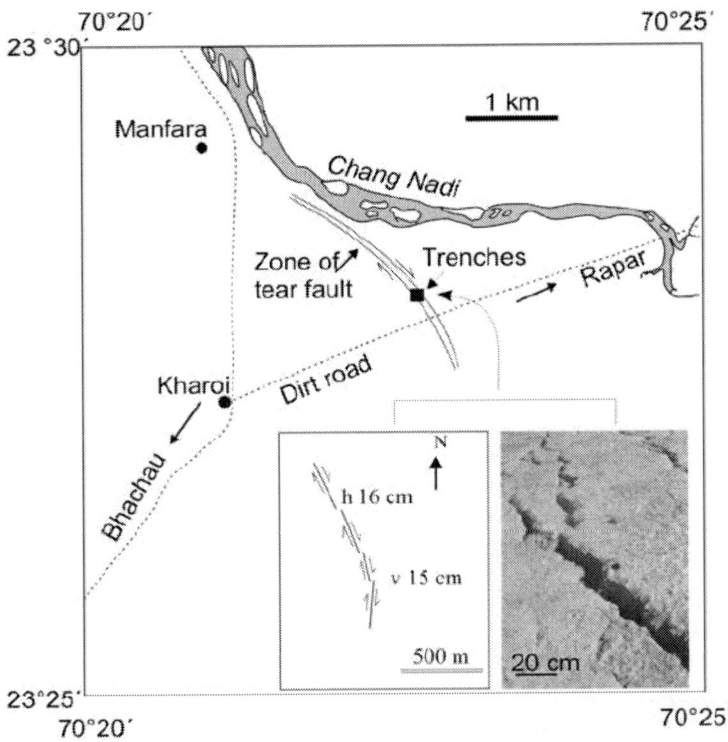

Figure 9. Location of Manfara fault and the site of the trenches. Inset: left- map view of the tear fault; right- view of the tear fault showing the right lateral offset with an oblique component.

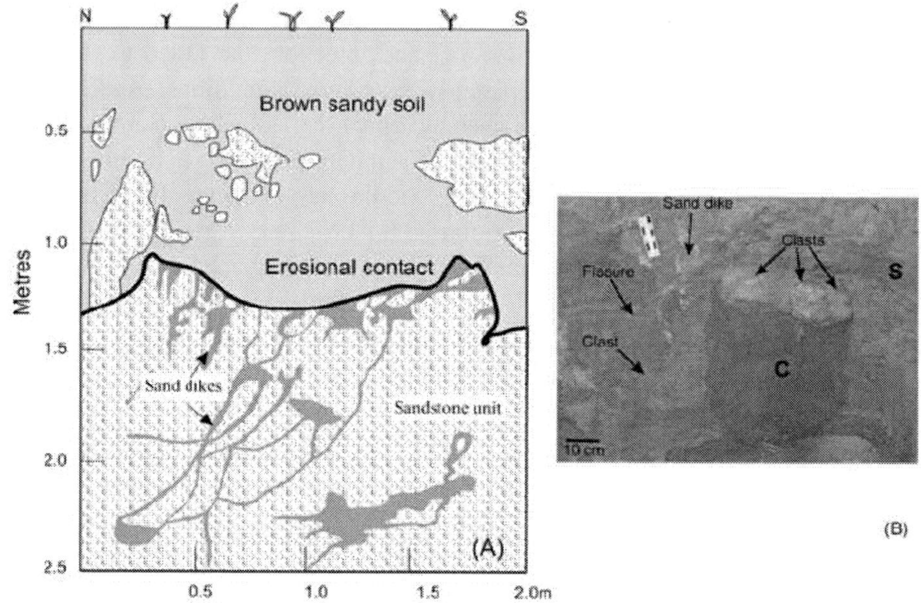

Figure 10. (A). Log of trench 1 exposing the pre-2001 generation of sand dikes. (B). Photograph of the lower part (1 m from the surface) of the section from trench 2; C- crater; S- sandy layer.

The 3-m-deep trench-2 excavated 20 m east of trench-1, revealed more evidence of earlier faulting events such as terminations of fault strands, vertical offsets of marker beds, colluvial wedges and fissure fillings (figure 11a and b), characteristic of regions of repeated earthquake activity [see Mc Calpin, 1996]. In this trench, the two sandstone units are separated by thin bed of gravel, defining a break in deposition and we consider the gravel layer as a paleo-surface. A third unit of sand has been deposited over the erosional contact above sands unit II.

We noted clear evidence of vertical displacement in the lower gravel layer with a maximum offset of ~20 cm (figure 11b). This offset has developed as a fissure that has been filled with clasts and pebbles derived from the gravel layer above, suggesting that the fissure fill followed the offset in the gravel horizon. Clearly, such fissures with fill derived from the top could have formed only if the surface was exposed and thus, we infer that the fissure was formed before sand unit II was deposited. The optically stimulated luminescence (OSL) date of the sample taken from the gravel bed is 5125±840 yr BP whereas the material from the fissure fill was dated at 4424±656 yr BP. The 2001 earthquake also led to similar fissures in this region but the displacement was only 10±5 cm. The 20 cm slip in the gravel bed may be cumulative, resulting from more than one earthquake. We assume that there must have been at least one earthquake prior to 2001 and based on the dates obtained from the gravel bed and the fissure fill, we assume that the OSL date of 4424±656 yr B.P to be the minimum age of the penultimate event.

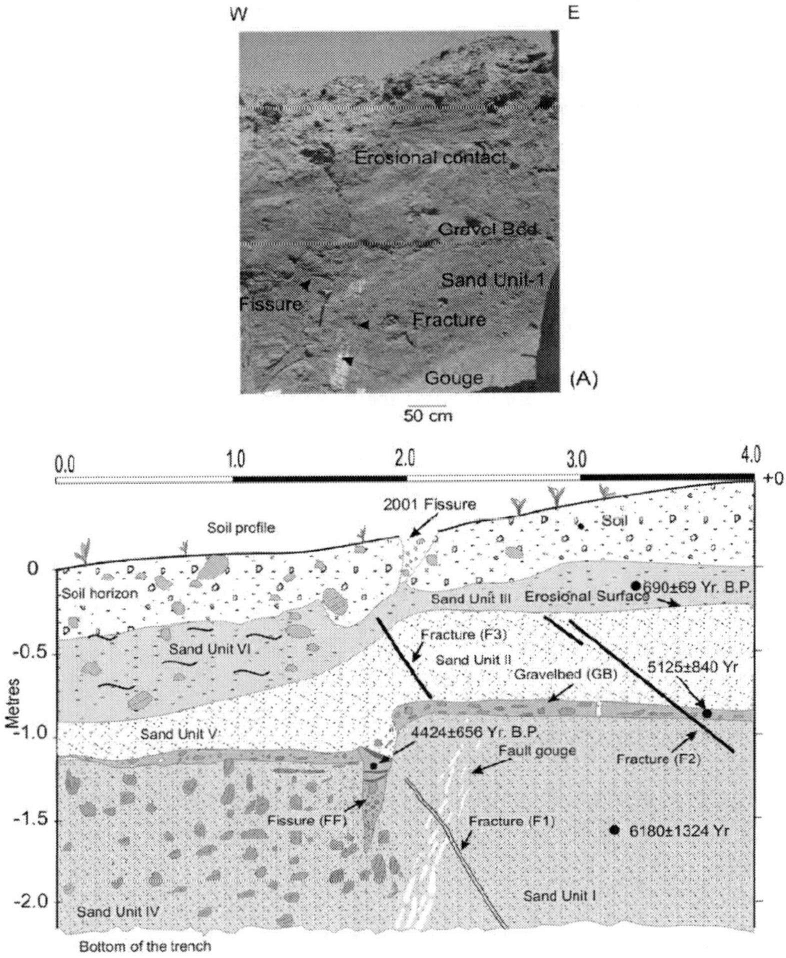

Figure 11. A. Photograph of the western wall of trench 3, excavated across the Manfara fault, showing the fault gouge at the base of the trench, an old fissure fill and fractures. B. Log of the western wall of trench 3, across the Manfara fault. The upthrown side corresponds to the footwall block and the downthrown side is the hanging wall.

We observed two sets of fractures on the trench walls, an older set (F1) terminating at the base of the sandstone unit I and another set that terminates at the top of the fine sand unit. A diffused and disturbed strand of whitish material occurs across the fracture that cuts the lower sandstone unit, a feature that we have observed in all the trenches across the Manfara fault. These are crushed grains of quartz and rock flour of siliceous material, generally indicative of frictional wear, and it occurs in a highly localized zone within the trend of the Manfara fault. X-ray diffraction analyses indicate that these patches of finer whitish rock flour are entirely composed of siliceous material without any clayey minerals. We interpret the ubiquitous presence of this gouge zone almost trending in the direction (north-south) of the Manfara fault as evidence for recurrent activity of crushing, shearing and consolidation.

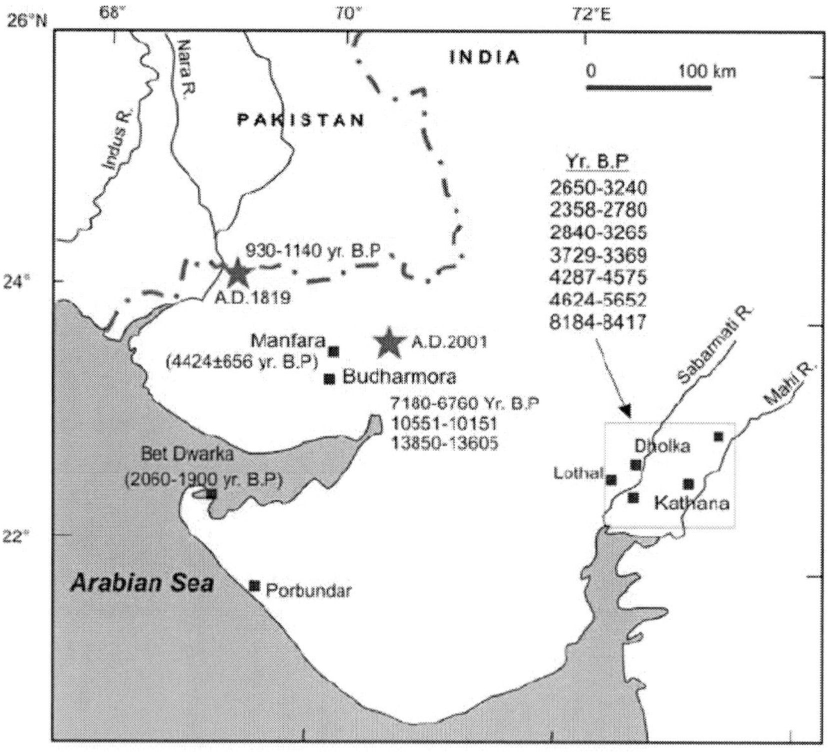

Figure 12. (a) Section excavated near Lothal (Uthalia) revealing three episodes of liquefaction (sandblows 1, 2 and 3), including the 2001 earthquake. Note the crosscutting relation of the 2001 liquefaction with the earlier event (sandblow-2).

Based on some of the known criteria which include the presence of a gouge zone, crack fill, upward termination of fractures and differential displacement at the fault plane [McCalpin, 1996], we consider this as a fault zone that is being reactivated from time to time. Stratigraphically, it is evident that at least one earlier event has disturbed the older sandstone unit and a long quiet interval must have passed before the occurrence of the 2001 event and therefore we consider the ~4000 year event as the penultimate event in this region. Rajendran et al. [2008] has presented results of trenching excavations in the Cambay basin and the distant effects of the event around 4000 yr BP seem to show up at more sites (see figure 12). Because of the fact that the 4000-year-old earthquake observed in the Manfara trench is associated with the faulting event at this site we assume that this could be the penultimate earthquake at the 2001 Bhuj source.

Earthquake Recurrence in the Kachchh Rift and Continental Seismicity

Large earthquakes in the stable continental interiors are few and are known to be associated with recurrence intervals, in the range of tens of thousands of years, making them appear 'out-of-the-blue' events, when viewed from the point of human life span. Much of the information on the past seismicity must be obtained through geological investigations and the results from known seismic source zones with in the continental interior regions suggest drastic variations in recurrence intervals and magnitudes among and within various SCRs. The largest and most of the moderate continental earthquakes have been associated with passive margins or failed rift systems [Johnston, 1996], the most prominent among them being the Charleston seismic zone (Southeastern US); New Madrid Seismic Zone (NMSZ, Central US) and the Kachchh seismic zone in NW India. Studies in the NMSZ and Charleston suggest that earthquakes of M >7 have recurrence interval of the order of hundreds of years [Amick and Gelinas, 1991; Tuttle and Schweig, 1995] and at least in the case of Charleston, the seismicity may be associated with multiple sources [Rajendran et al., 1993]. Earthquakes in the unrifted continental regions, such as those in the Australian continent are moderate and are suggested to be associated with recurrence interval of the order of several thousands of years [Crone et al., 1997]. The available data suggest that it may be possible to tentatively group the SCR source zones on the basis of their relationship to their tectonic environments, but the recurrence pattern remains less understood, mostly due to the rarity of such events.

In the above context, the earthquakes in the Kachchh zone provide an unprecedented opportunity because no other site of continental seismicity, has generated large earthquakes in the recent times, offering the opportunity to use the coseismic features as proxies for identifying previous events. Had the 2001 earthquake not occurred, we would have remained oblivious of other seismogenic sources capable of generating large earthquakes in the Kachchh rift. The post-2001 exploration led not only to the identification of sources independent of the 1819 and 2001, but also suggested sources with varying interseismic periods. The available chronological constraints do not permit suggestion of a possible recurrence interval. Based on the 1819 and its penultimate earthquake about 1000 years ago, we suggest that the Allah Bund fault, the well-defined boundary fault of the Kachchh rift, may generate earthquakes at an interval of about 1000 years. But such a relatively shorter recurrence cannot be attributed to the 2001 source, which seems to have generated one earthquake about 4000 yr ago (as evident from Manfara) and another one around 7000 yr ago as evident from Budharmora. The ~2000-year-old event from Bet Dwarka and ~3000 yr BP event observed at Dholka are probably related to independent sources. In summary, based on the work do far, we conclude that the Kachchh rift as well as the Cambay rift hosts multiple sources, which are being reactivated at varying intervals.

Subsequent to the 2001 earthquake there has been renewed discussion on continental seismicity, and often the Kachchh seismic zone has been compared with the New Madrid seismic zone in the eastern United States [Schweig and Ellis, 1994]. Stein et al. [2002], however, suggest that the Kachchh seismic zone, which is only 400 km away from the diffuse plate boundary between the Indian and Eurasian plates may be referred to more as an "interplate" site rather than SCR. Stein et al. [2002] further point out that the New Madrid seismic zone is located ~2400 km from the plate boundary. Perhaps what is common between these two regions is that they are associated with the rift related tectonics. The fault plane solutions of earthquakes in the Kachchh rift suggest reverse faulting on E-W oriented fault planes, both for the 1956 Anjar earthquake [Chung and Gao, 1995] as well as the 2001 Bhuj earthquake [Mandal et al., 2004]. The relatively low level seismicity between the Kachchh seismic zone and the plate boundary to the northwest, and the fact that the plate boundary to the northwest is characterized by predominantly left-lateral strike slip faults (see figures 2 and 4 of Stein et al., 2002] suggest that these belong to different tectonic provinces.

The studies in the New Madrid seismic zone suggests intervals of ~ 200 – 500 years for large earthquakes [Tuttle and Schweig, 1996]. However, due to large spatial extent of liquefaction features, it is not easy to identify the source of these earthquakes. It is not evident if the earthquakes originate from the same source zone or if they are related to multiple sources. The observations in the Kachchh seismic zone suggest existence of multiple sources associated with varying recurrence intervals. The Narmada rift, a Precambrian rift system that cuts through the Indian peninsula is also known to be associated with mid and lower crustal, moderate earthquakes, the last being the MW 5.7 earthquake in 1997. It has been suggested that the earthquakes in the Narmada rift are related to the lower crustal fossil rift pillows [Rajendran and Rajendran, 1998; 1999; Rao et al., 2002] similar to the suggested mechanism for the lower crustal, rift-related earthquakes in the Amazona rift [Zoback and Richardson, 1996]. Pollitz et al. [2001] have suggested a similar mechanism for the New Madrid seismic zone. Compared to the New Madrid seismic zone that is spatially well-defined by the recent seismicity, the Kachchh seismic zone has been very quiet during the intervening periods. None of the major fault systems, including the Allah Bund fault and the causative fault of the 2001 earthquake are expressed by any background low-level seismicity.

Investigations in the Kachchh seismic zone suggest that the seismic activity within a rift zone can be associated either with the boundary faults or with the emergent faults within the rift. Studies in this region also suggest that the style of deformation, recurrence pattern and nature of coseismic features can vary within the same source zone.

Conclusion

Investigations in the Kachchh rift region have helped add new information on the major earthquakes of the region during the last 10K years. At least one of these earthquakes occurred about 1000 yr BP seems to be associated with the Allah Bund fault, coincident with the northern boundary of the Kachchh rift. The date of this event and the likely location has been inferred from geologic data, and also corroborated from historic documents. The 2001 earthquake generated a variety of coseismic features that could be used as proxies of older events, but the lack of primary fault rupture precluded estimate of coseismic and penultimate slip on the causative fault. However, the secondary faulting, lateral spreads and sandblows generated in the deformation zone offered possibilities to search for similar features formed in the past. Features include dextural tear faults formed near Manfara and the lateral spread at Budharmora to understand the previous activity in the epicentral zone of the 2001 Bhuj earthquake.

The study at Manfara demonstrates the usefulness of secondary faults in the studies of earthquake histories where the primary fault remains blind. Archival records of an earlier vertical displacement exposed in the trenches at Manfara suggest that the previous earthquake in this source may have occurred about 4000 yr ago. The tear fault at Manfara generated by the 2001 earthquake must be part of a transverse fault linking the Kachchh Mainland and South Wagad faults [Biswas and Khattri, 2002]. Evidence for at least one earthquake observed in the trenches supports the "slip on the pre-existing structure" hypothesis proposed by McCalpin and Thakkar [2003] to explain the mechanism of secondary faulting associated with the 2001 earthquake.

The spatial extent of the 2001 coseismic features has provided some useful indications to identify the potential sources of ancient earthquakes. For example, Rajendran and Rajendran [2003] suggest that no major earthquakes have sourced at this region in the historic past. The feeble earthquake-induced features and low-intensity damage to the archeological site at Dholavira (a settlement that lasted about ~1500 years, during 3450-4950 yr BP) during the 2001 earthquake was the basis of their argument. The 1819 earthquake, the strongest known historical event in this region also did not produce any significant effects in this region [see Oldham, 1926]. However, archeological evidence suggest that an earthquake caused minor damage to this settlement during 4150-4450 yr BP [Joshi and Bisht, 1994]. The 2001 earthquake, located 60 km to its south, caused minor damage to the monument at Dholavira, like the collapse of the pillar at the entrance.

We observed several coseismically induced liquefaction features in the region and also moderate damage in the village in the vicinity of ancient Dholavira settlement. Thus it appears that the earthquakes from the 1819 source have not affected this region whereas those from the 2001 or nearby sources have a larger impact in this region. Thus, it is quite possible that the reported damage during 4150-4450 yr BP, if related to an earthquake, could have originated from a distant source, possibly closer to the 2001 earthquake. Evidence for a contemporaneous earthquake has been reported from our work at Budharmora and Lothal. The present level of understanding leads us to conclude that the 4000-year-old earthquake could have originated either from the 2001 or at a nearby unknown source. Evidence for the event prior to the 4000-year-event is sketchy; the Budharmora trenches suggesting an event around 9000 yr BP. We suggest that the recurrence interval of earthquakes vary spatially within the Kachchh rift. The 1819 source may be associated with ~1000 yr interval whereas at the 2001 source, the recurrence interval may be much longer, closer to 4000 yr BP. The earthquake history derived from paleoseismological data, not complete by any means, provides new insights into the temporal and spatial characteristics of the Kachchh seismicity.

A close look at the global pattern of continental earthquakes suggests that the intracontinental region of India is considerably more active. Proximity to an active plate boundary may be pointed out as one factor that makes this region different from its counterparts. The seismicity associated with the Kachchh rift has been linked to the diffuse plate boundary ~400 km away [Stein et al., 2002]. However, the very low-level seismicity associated with the neighboring Arabian Peninsula [see figure 1 of Stein et al., 2002] in the close proximity of a plate boundary is quite striking. The low P-wave velocity anomaly and the higher heat flow in the northern portion of the Cambay graben have been directly linked with the Deccan plume activity [Kennett and Widiyantoro, 1999]. More recent studies based on S-receiver functions by Kumar et al. [2008] suggest that the Indian lithosphere is comparatively much thinner (~100 km), as a consequence of which it moves faster. The clustered nature of the immediate aftershock activity [Bodin and Horton, 2004] mid-crustal source of the main shock and the low velocity zone associated with the 2001 source zone [Mandal, 2008] suggest that the midcrustal rift pillow could act as a source of stress concentration. Could this mechanism explain the existence of multiple sources? Their temporally independent behavior is another issue that future investigations must consider. The present paleoseismological investigations have helped to isolate some of the possible sources and provide a template, the future studies must focus on developing better constraints of their characteristics.

ACKNOWLEDGMENTS

The Department of Science and Technology (DST), Government of India supported most of our studies in the Rann of Kachchh, since 1998, and part of the work presented in this paper was also partially funded by the National Institute of Ocean Technology. We also thank DST for awarding Ramanujan Fellowship to the first author of this article. Numerous people have helped us during the execution of this work, we thank them all, in particular the Indian Border Security Force.

REFERENCES

Amick, D.; Gelinas, R. *Science.* 1991, 251, 655-658.

Antoliok, M; Dreger, D. S. *Bull. Seismol. Soc. Am.* 2003, 93, 1235-1248.

Bilham, R. In Stewart, I. S.: Vita-Finzi, C.; Ed.; *Coastal Tectonics,* 146, Geological Society: London, 1999. 295-318.

Biswas, S. K.; Khattri, K. N. *J. Geol. Soc. India.* 2002, 60, 131-142.

Bodin, P.; Horton, S. *Bull. Seismol. Soc. Am.* 2004, 94, 818-827.

Crone, A. J.; Machette, M. N.; Bowman, J. R. *Aust. J. Ear. Sci.* 1997, 44, 203-214

Chung, W. P.; Gao, H. *Tectonophysics.* 1995, 42, 281-292.

Schweig, E. S.; Ellis, M. *Science*. 1994, 264, 1308-1311.

Gahalaut, V. K.; Burgmann, R. Bull. Seism. Soc. Am. 2004, vol. 94, 2407-2413.

Gaur, A. S.; Vora, K. H. *Cur. Sci.* 1999, 77, 180-185.

Grant, L. B.; Sieh, K. J. *Geophys. Res.* 1994, vol. 99, 6819-6841.

Iyengar, R. N.; Sharma, D.; Siddqui, J.M. *Indian Jour. Hist. Sci.,* 1999, vol. 34, 181-237.

Johnston, A. C. *Geophys. Jour. Int.* 1996, 125, 639-678.

Joshi, V. P.; Bisht, R.S. India and the Indus Civilization, National Museum Institute, New Delhi, 1994, pp 22-31.

Kayal, J. R.; Zhao, D.; Mishra, O.P.; Singh, O.P. *Geophys. Res. Lett.* 2002, 29, doi: 10, 1029/2002GL015177.

Kumar, P.; Yuan, X.; Kumar, R.; Kind, R.; Li, X.; Chadha, R. K. *Nature*. 2008, 449, 894-897.

Lyell, C.; P*rinciples of Geology,* Appleton & Co. New York, 1857. 834pp.

Mac Murdo, *J. Phil. Mag.*, 1824, 63, 105-177.

Mandal, P.; Rastogi, B. K.; Satyanarayana, V. H. S.; Kousalya, M.; Vijayaraghavan, R.; Satyamurty,C.; Raju, I. P.; Sarma, A. N. S.; Kumar, N. T*ectonophysics*. 2004, 378, 105-121.

Mandal, P. *PAGEOPH.* 2008,165, 1307-1324.

Maurya, D. M.; Rachana, R.; Chamyal, L. S. *Jour. Geol. Soc. India.* 1998, 51, 755-758.

McCalpin, J. P.; Ed.; Paleoseismology, Academic Press: New York, 1996, 588pp.

McCalpin, J. P.; Thakkar, M. G. *Ann. Geophys.* 2003, 46. 937-956.

McCalpin, J. P.; Zuchiewicz, W.; Jones, L. C. *Jour. Sediment. Pet.* 1993, 63, 120-130.

Meghraoui, M.; Delouis, B.; Ferry, M.; Giardini, D.; Huggenberger, P.; Spottke, I.; Granet, M. *Science.* 2001, 293, 2070-2073.

Newman, A.; Stein, S.; Weber, J.; Engeln, J.; Mao, A.; Dixon, T. *Science.* 1999, 284, 619-621.

Noller, J. S.; Sowers, J. M.; Lettis, W. R. *Quaternary geochronological methods and applications*; American Geophysical Union: Washington, DC, 2000; 588pp.

Obermeier, S. F. Using liquefaction induced features for paleoseismic analysis, *U. S. Geol. Surv. Open File Rept.,* 1994; 94-663, 58pp.

Obermeier, S. F. 1996. In *Paleoseismology;* McCalpin, J.; Ed.; Academic Press: New York, 1996, pp. 331-396.

Oldham, R. D. The Cutch (Kachh) earthquake of the 16th June, 1819 with a revision of the great earthquake of the 12th June, 1897, India, *Geological Survey Memoir,* Calcutta, India, 1926; Vol. 46 (Part 2), 77pp.

Pollitz, F.; Kellogg, L.; Burgmann, R. *Bull. Seism. Soc. Am.* 2001, 91, 1882-1897.

Rajendran, C. P.; Talwani, P. *Geology.* 1993, vol. 21, 987-990.

Rajendran, C. P.; Rajendran, K.; John, B. 1996, 24, 651-654.

Rajendran, K.; Rajendran, C. P. *Cur. Sci.* 1998. 24, 168-174.

Rajendran, K.; Rajendran, C. P. *Tectonophysics.* 1999, 305, 355-370.

Rajendran, C. P.; Rajendran, K. B*ull. Seism. Soc. Am.* 2001. 91, 3, 407-426.

Rajendran, K.; Rajendran, C.P.; Thakkar, M.; Tuttle, M.P. *Cur. Sci.* 2001, 80, 1397-1405.

Rajendran, K.; Rajendran, C.P., Thakkar, M.G.; Gartia, R.K. *Cur. Sci.*, 2002, 83, 603-610.

Rajendran, C. P.; Rajendran, K. *Seism. Res. Lett.*, 2002, 73, 470-479.

Rajendran, C. P.; Rajendran, K. In *Memoir,* Geological Society of India, Bangalore, 2003a, 54, 87-142.

Rajendran, C. P.; Rajendran, K. Studying earthquake recurrence in the Kachchh region, India. *EOS*. 2003b, 84, 529, 533, 536.

Rajendran, C. P.**;** Rajendran, K.; Vora, K.H.; Gaur, A.S. *Cur. Sci.,* 2003, 84, 695-701.

Rajendran, C. P.; Rajendran, K.; Thakkar, M.; Goyal, B. J. *Geophys. Res.*, 2008, 113, doi:10.1029/2006JB004845.

Rao, P. N.; Sukuda, T.; Kosuga, M.; Bhatia, S. C.; Suresh, G. *Geophys. J. Internat.*, 2002, 148, 132-138.

Schwartz, D. P.; Coppersmith, K. J. *J. Geophys. Res.,* 1984, 89, 5681-98.

Schweig, E. S,; Gomberg, J.; Petersen, M.; Ellis, M.; Bodin, P.; Mayrose, L.; Rastogi, B. K. *Geol. Soc. India,* 2003. 61, 277-282.

Sieh, K. E. *J. Geophys. Res.,* 1978, 83, 3907-3939.

Sims, J. D.; Garvin, C. D. *Bull. Seismol. Soc. Am.*, 1995, 85, 51–65.

Singh, S. K.; Pacheco, J. F.; Bansal, B. K.; Perez-Campos; Dattatrayam, R. S.; Suresh, G., *Bull. Seismol. Soc. Am.,* 2004, 94, 1195-1206.

Stein, S.; Sella, G.; Okal, E. A. In *Plate Boundary Zones*, Stein, S.; Freymueller, J. T.; Ed.; American Geophysical Union: 2002, pp 243-254.

Talwani, P.; Gangopadhyay, A.; 2001, *Seism. Res. Lett.,* 72, 336-345.

Talwani, P.; Schaeffer, W. T. *J. Geophys. Res.* 2001, 106, 6621-6642.

Tuttle, M. P.; Schweig, E. S. *Geology*. 1991, 23, 253-256.

Tuttle, M. P.; Schweig, E. S.; Sims, J. D.; Lafferty, R. H.; Wolf, L. W.; Haynes, M.L. *Bull. Seismol. Soc. Am.* 2002, 92, 2080--2089.

Tuttle, M. P.; Hengesh, J.; Tucker, K. B.; Lettis, W.; Deaton, S. L.; Frost, J. D. *Earthquake Spectra.* 2002, vol. 18 (suppl. A), 79-100.

Vora, K. H.; Gaur, A. S.; Price, D.; Sundaresh, *Cur. Sci.* 2002, vol. 82, 1351-1356.

Wesnounsky, S. G.; Seeber, L.; Rockwell, T. K.; Thakur, V. C.; Briggs, S.; Kumar, S.; Ragona, D. *Seism. Res. Lett.*, 2001, 72, 514-524.

Yeats, R. S.; Sieh, K.; Allen, C. R. The *Geology of Earthquakes,* Oxford University Press: New York, 1997. 568 pp.

Zoback, M.L.; Richardson, R.M. *J. Geophys. Res.* 1996, 101, 5459-5475.

In: Recent Progress on Earthquake Geology ISBN: 978-1-60876-147-0
Editor: Pierpaolo Guarnieri, pp. 49-78

Chapter 3

EVIDENCE OF ACTIVE TECTONICS IN SOUTHERN ITALY: THE ROSSANO FAULT (CALABRIA)

***P. Galli*[1], *V. Spina*[2], *I. Ilardo*[3] *and G. Naso*[1]**

[1] Dipartimento Protezione Civile, U. Rischio Sismico, Rome, Italy.
[2] Total Exploration and Production, Rome, Italy.
[3] Free-lance geologist, Lascari (Pa), Italy.

ABSTRACT

Unlike the Tyrrhenian side, the seismotectonics of the Ionian sector of the Calabrian peninsula is still not well defined, both in terms of active tectonics and seismic activity. This work focuses on a strong earthquake that occurred along the northern Ionian coast of Calabria in 1836 (Rossano area, Mw=6.2), by applying a multidisciplinary approach in order to find out its possible causative fault. Thanks to the reappraisal of the historical accounts of the earthquake - which also contain possible surface break descriptions - and to geological, geomorphological, structural and paleoseismological analyses carried out in the mesoseismic area, we propose that the poorly known, E-W striking, normal Rossano Fault might have been responsible for the 1836 event. This fault, which is characterized by an impressive ~100-m-high rectilinear scarp, downthrows the Palaeozoic rocks of the Sila Massif below the coastal alluvial plain of Rossano. Absolute dating of faulted talus-deposits, coupled with observations on the geometry and aspect of the fault free-face, allow to hypothesize repeated surface ruptures in the Late Holocene.

Keywords: Active fault, paleoseismology, Calabria, 1836 earthquake.

1. INTRODUCTION

The Calabrian peninsula is the arched narrow tip of the Italian "boot", bridging the NW-SE Apennine chain and the roughly opposite ~E-W striking Siculo-Maghrebides chain. Its western side, from the Crati Basin to the Messina Straits, is characterized by one of the strongest seismicity in the Mediterranean, with earthquakes exceeding magnitude 7 (figure 1). Most of these earthquakes are caused by the extensional faults that downthrow the Palaeozoic massifs of Sila, Serre and Aspromonte (i.e., the crystalline backbone of the Calabrian Arc) toward the Tyrrhenian Sea (Ghisetti, 1981; Tortorici et al., 1995), as confirmed definitely by paleoseismic studies carried out across the Cittanova fault (9 in figure 1: ~372 AD earthquakes, and February 5, 1783 earthquake, Mw~7; Galli and Bosi, 2002), or the S-Serre fault (8 in figure 1: February 7, 1783 earthquake, Mw~6.7; Galli et al., 2007). However, apart from strong events reasonably related to similar structures (e.g., the one responsible for the catastrophic 1908 Messina earthquake, Mw=7.3; see insights in Argnani et al., 2008), or others related to oblique faulting in the inner Sila massif (e.g., Lakes Fault, 5 in figure 1: June 9, 1638 earthquake, Mw~6.8; Galli and Bosi, 2003; Galli et al., 2007), many others have not been still associated to a certain seismogenic source. This is particularly true for the few earthquakes known to have occurred on the Ionian side of Calabria, such as the 1832 (Mw~6.6) and the 1836 (Mw~6.2) events (figure 1; see insights and references in Galli et al., 2007).

In this work we focus on the area of the 1836 earthquake, where we investigated the poorly known normal Rossano Fault (RF from now). Firstly, we carried out a reappraisal of the effects of this earthquake on the basis of all the available coeval documents, some of which found in local archives. Then, through the interpretation of aerial photographs and field survey, we tried to piece together the overall Quaternary evolution of the area, finally identifying a site across the RF, which was prone to the excavation of an explorative paleoseismological trench. The comparison between historical data (e.g. informations on surficial breaks and the highest intensity datapoint distribution) and field results (geological, geomorphological, structural and paleoseismological data) allowed us to hypothesize repeated ruptures of this structure in the Late Holocene, the last one presumably related to the 1836 earthquake.

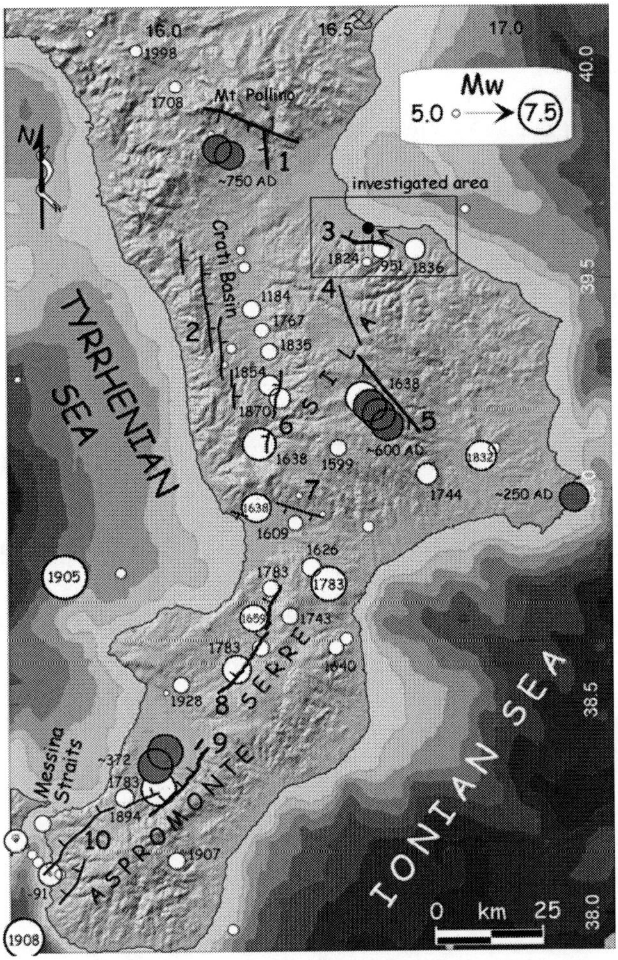

Figure 1. Earthquakes distribution in Calabria (mod. from Working Group CPTI, 2004. Mw>5.5) and primary seismogenetic faults (bold=certain, i.e. investigated by paleoseismological analyses). 1) Mt. Pollino Fault; 2) W-Crati Fault System; 3) Rossano Fault (this work); 4) Cecita Fault; 5) Lakes Fault; 6) Savuto Fault System (Piano Lago-Decollatura Fault); 7) Feroleto-Sant'Eufemia Fault; 8) Serre Fault System; 9) Cittanova Fault; 10) Reggio Calabria Fault System. Shaded earthquakes symbols derive from paleoseismic and/or archaeoseismic studies (see in Galli et al., 2008). The rectangle shows the investigated area, where the black circle evidences our preferred 1836 earthquake epicentre (i.e., with respect to Working Group CPTI, 2004).

2. Overview of the Local Seismicity

Calabria is struck by the strongest earthquakes of the whole Apennine chain (figure 1). Examples include the 1905 (Ms=7.5) and 1908 (Ms=7.3) earthquakes, and the catastrophic sequences of March-June 1638 (Mw~6.6÷6.8) and February-March 1783 (Mw~6.6÷7.0). Most of the Calabrian records of seismicity are clustered between the 17th century and the beginning of the 20th century, whereas we known only sparse earthquakes during the Middle Age (e.g., ~951 in the Rossano area; 1184 in the Crati Valley), and few others discovered through paleoseismological investigations (see Galli et al., 2008 and reference therein). As suggested by Scionti et al. (2006), this seismic pattern during the past millennium may be due to the seismogenic behaviour of the primary Calabrian active faults, which triggered each other starting from the early 17th cent., and not to the loss of archive data that, reasonably, involved only M<6.5 events information (e.g. in Stucchi and Albini, 2000).

Actually, few local earthquakes are on record in the Rossano area; the oldest known happened in ~951 AD and - according to *Bartholomaeus Rossanensis* (11th cent.) - it rocked the Rossano town, causing strong damage and landslides. In 1824, a moderate event caused light damage in Rossano and Corigliano, anticipating the strong earthquake of April 25, 1836 (Io=9 MCS, Mw=6.2, Working Group CPTI, 2004; from now CPTI). Recently (December 2, 1995), a small event (Ml=3.4) occurred close to Rossano; its focal mechanism suggests normal faulting on roughly E-W planes (Vannucci and Gasperini, 2004; figure 2).

2.1. The 1836 Rossano Earthquake

On the basis of all the contemporary sources - some of which are unpublished manuscripts that we read in the Rossano municipality archive - we re-evaluated the effects distribution of the 1836 earthquake (results are summarized in table 1 and in figure 2). The strongest effects occurred in the town of Rossano (9-10 MCS; Mercalli-Cancani-Sieberg scale), and in the small villages of Crosia (10 MCS) and Calopezzati (9 MCS), whereas the area of severe damage (i.e., 7 MCS) extends ~70 km in a WNW-ESE direction, enveloping about 20 villages.

Microtremor analyses that we carried out in all the localities of the mesoseismic area (following the Nakamura [1989] method; see some examples in figure 3) do not show HVSR peaks in the engineering range (2-5 Hz), suggesting that none of villages suffered particular geological site amplification.

Table 1. MCS intensity distribution re-evaluated for the 1836 earthquake

LOCALITY	LON	LAT	I (MCS)
Crosia	16,773	39,566	10
Rossano	16,635	39,574	9-10
Calopezzati	16,802	39,56	9
Caloveto	16,76	39,504	8-9
Corigliano calabro	16,518	39,596	8
Cropalati	16,726	39,516	8
Paludi	16,682	39,529	8
Schiavonea	16,541	39,652	8
Torre Cilento	16,538	39,6259	8
Scala Coeli	16,889	39,445	7-8
Bocchigliero	16,751	39,418	7-8
Campana	16,824	39,411	7-8
Cariati	16,949	39,497	7-8
Longobucco	16,611	39,449	7-8
Acri	16,386	39,49	7
Bisignano	16,285	39,513	7
San Demetrio Corone	16,362	39,568	7
San Giorgio Albanese	16,454	39,582	7
Castiglione Cosentino	16,288	39,351	7
Craco	16,439	40,377	7
Cosenza	16,251	39,303	6
Donnici sup.	16,301	39,251	6
Cirò	17,064	39,38	6
Rose	16,288	39,398	6
Motta	16,314	39,304	6
Rovito	16,321	39,308	6
San Pietro Guarano	16,311	39,342	6
Rovella	16,288	39,303	6
Zumpano	16,291	39,31	6
Ginosa	16,758	40,578	6
Castrovillari	16,202	39,814	5
Crotone	17,127	39,08	5
San Mauro	16,925	39,103	5
Santa Severina	16,913	39,146	5
Scandale	16,959	39,121	5
Strongoli	17,05	39,265	5
Catanzaro	16,586	38,914	5
Spezzano	16,3108	39,669	5
Nicastro	16,318	38,974	4-5

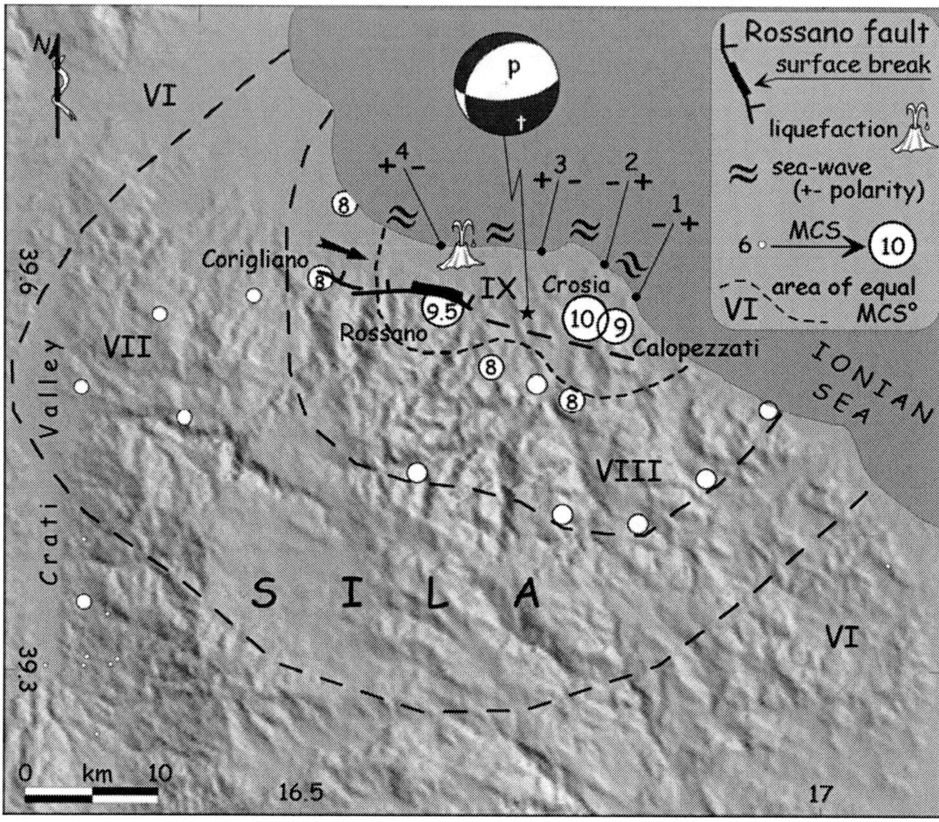

Figure 2. Intensity distribution of the 1836 earthquake. Note that the isoseismic lines are opened seward, and that the mesoseismic area roughly parallels the Rossano Fault; this fact suggests that this fault could be associated with the 1836 seismogenic structure. The *tsunamis* effects along the coast (site 1, Caloveto beach; 2, Centofontane; 3, Iapichello; 4, Sant'Angelo; from Rossi, 1836; De Rosis, 1838) imply the occurrence of a coseismic deformation of the area; polarity might indicate the lowering in the nearfield of the Rossano Fault hangingwall (positive, sea ingression), with sea retreat in the surrounding zones (negative). The black arrow indicates the possible fault rupture direction, accounting for strong effects in the small villages of Crosia and Calopezzati. The focal mechanism refers to the Ml=3.4 earthquake of December 02, 1995 (Vannucci and Gasperini, 2004; p, P-axis; t, T-axis).

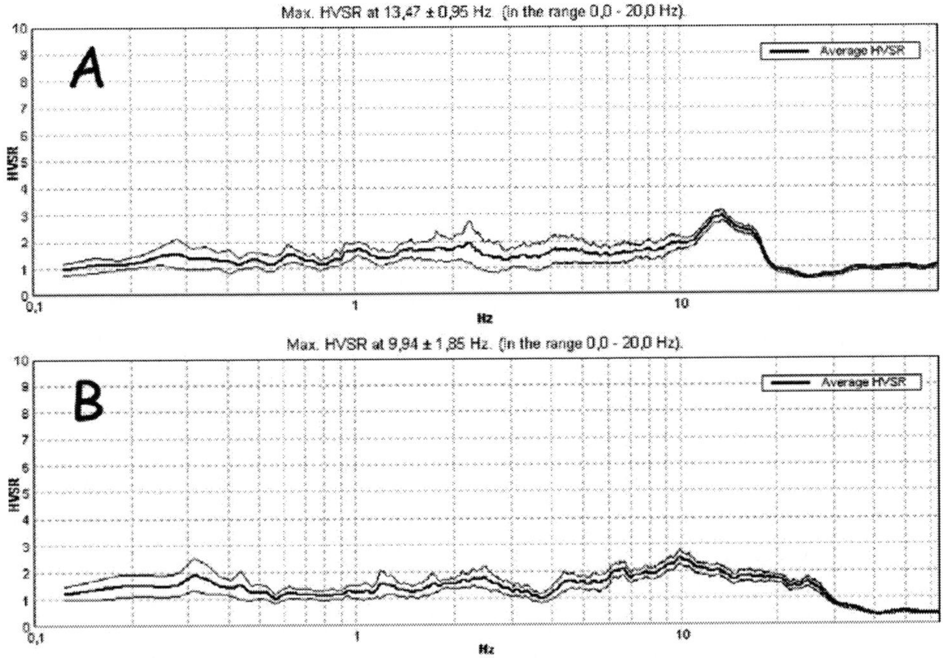

Figure 3. HVSR obtained in Rossano (A) and Crosia (B). Note the absence of amplification for any frequency. This fact has been observed in all the villages where we performed microtremor analyses.

On the other hand, as observable in figure 2, the 1836 highest intensity datapoint distribution (HIDD) is clearly "opened" seaward, its barycentre (i.e., its rough hypocenter) falling along the Rossano-Crosia coast. Here, along the shoreline, contemporary sources (i.e., Rossi, 1836; De Rosis, 1838) described an abrupt, coseismic sea-retreat, followed by a violent sea-wave (sites 1-2 in figure 2), and *viceversa* (sites 3-4). We hypothesize that the different polarity of the sea-wave might indicate the coseismic lowering of the RF hangingwall (i.e., positive = sea ingression), whereas surrounding zones recorded a first negative impulse (= sea retreat). Moreover, along the Rossano coast plain, liquefaction phenomena were also observed (figure 2; open vents with ejection of water and sand and sand-volcanoes; see in Galli, 2000 and references therein), whereas significant surface breaks were described along the hillside of Rossano (Romano, 1836; Rossi, 1836; De Rosis, 1838). In particular, some of these breaks were described as hundred-meters long, and as cutting through both sides of the valleys. These features fit with the RF trace (see location in figure 2), and thus we suggest that they could be ascribed to surface faulting phenomena.

3. Geological Framework of Northern Calabria and of the Investigated Area

The Calabrian Arc is built of Palaeozoic crystalline-metamorphic rocks (Calabride Complex, *sensu* Amodio Morelli et al., 1976) belonging to the European-Iblean margin. Between 35 and 30 Ma the Neo-Tethyan slab-rollback induced the SE-translation of the orogenic front (Jolivet and Faccenna, 2000). As a result of that, since the Latest Oligocene, the Calabrian block thrusted onto ophiolite-bearing units and, since Middle Miocene, this thrust-nappe edifice overthrusted the Mesozoic carbonates of the African domain (Amodio Morelli et al., 1976; Dewey et al., 1989; Messina et al., 1991; Rossetti et al., 2001).

According to Van Dijk et al., (2000), the main Neogene tectonic structures of the northern Calabrian Arc consist of NW-SE sinistral strike-slip crustal shear zones (e.g. Pollino and S. Nicola-Rossano Fault zones; PFZ and SNRFZ in figure 4), with associate thrusts in their restraining bends; these structures controlled both the architecture and the tectono-sedimentary evolution of the Neogene basins, such as the Rossano and Crotone Basins (figure 4; Van Dijk & Okkes, 1991). All the tectonic units deriving from the structuring of the orogen form the core of the Calabrian Arc (e.g. the Sila Massif), and are covered by thick sedimentary successions which fill complex basins (e.g., Crati, Rossano, Crotone basins; see figure 4).

In the investigated area, northward of the Sila Massif slopes (figures 4 and 5) the Rossano basin is filled by a late Miocene succession, composed of sands, calcarenites and conglomerates, alternating with clays. This succession evolves upward to clays with gypsum, ending with terrigenous deposits (Roda, 1964; thickness of ~1300 m in borehole Trionto-1, available at internet site: http://www.socgeol.info/pozzi/index.asp; location in figure 5). The Miocene sequence is unconformably covered by Plio-Quaternary sediments (figures 4 and 6), composed of basal fluvial-deltaic conglomerates, sands and marine gravels.

The Pleistocene units evolve both laterally and vertically to silty clays (Roda, 1964, Vezzani, 1968, Carobene, 2003) which, in turn, contain tephra levels dated ~450 ka by Bigazzi and Carobene (2004). During Middle Pleistocene (i.e., after ~450 ka) the area experienced strong regional uplift, as testified by the wide extension of outcropping deltaic and alluvial systems, and by the impressive staircase of alluvial and marine terraces.

The entire Neogene succession plunges northward (figures 6, 7 and 8), with strata steepening on nearing the Sila Massif bedrock, and it is cut by ~E-W normal faults. The most important - in terms of both offset and length - is the N-dipping RF (figure 9). As it will be discussed later, the RF has a ~12-km-long surficial expression, and it is mainly arranged in a series of overlapping and *en-échelon* fault segments, the longest of which are the Corigliano and Rossano segments (labelled with a and b in figure 5).

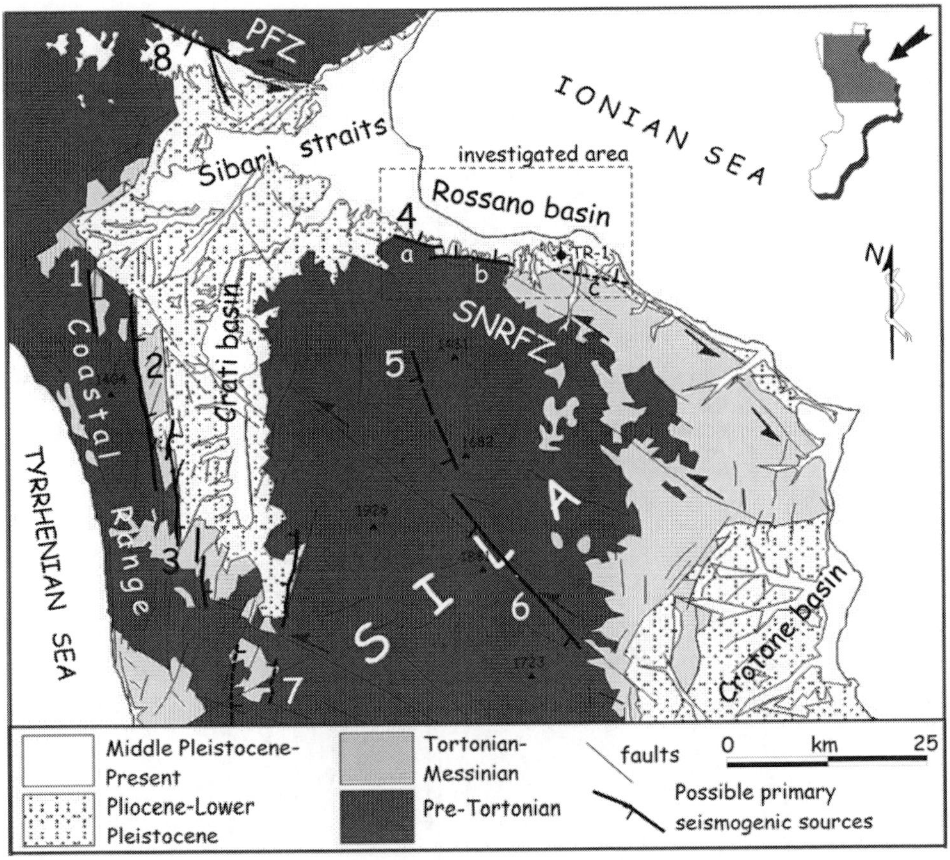

Figure 4. Structural sketch of northern Calabria and main possible seismogenetic faults: 1-3, W-Crati Fault System (1, Fagnano-Castello; 2, San Marco-San Fili; 3, Montaldo-Rende); 4, Rossano (segment a: Corigliano; b, Rossano; c, Crosia); 5, Cecita Lake; 6, Lakes; 7, Piano Lago-Decollatura; 8, Mt. Pollino. PFZ (Pollino Fault zone) and SNRFZ (S. Nicola-Rossano Fault zone) represent crustal shear zone. Only faults 4, 6, and 8 have been studied by means of paleoseismological analyses. TR-1, Trionto 1 borehole.

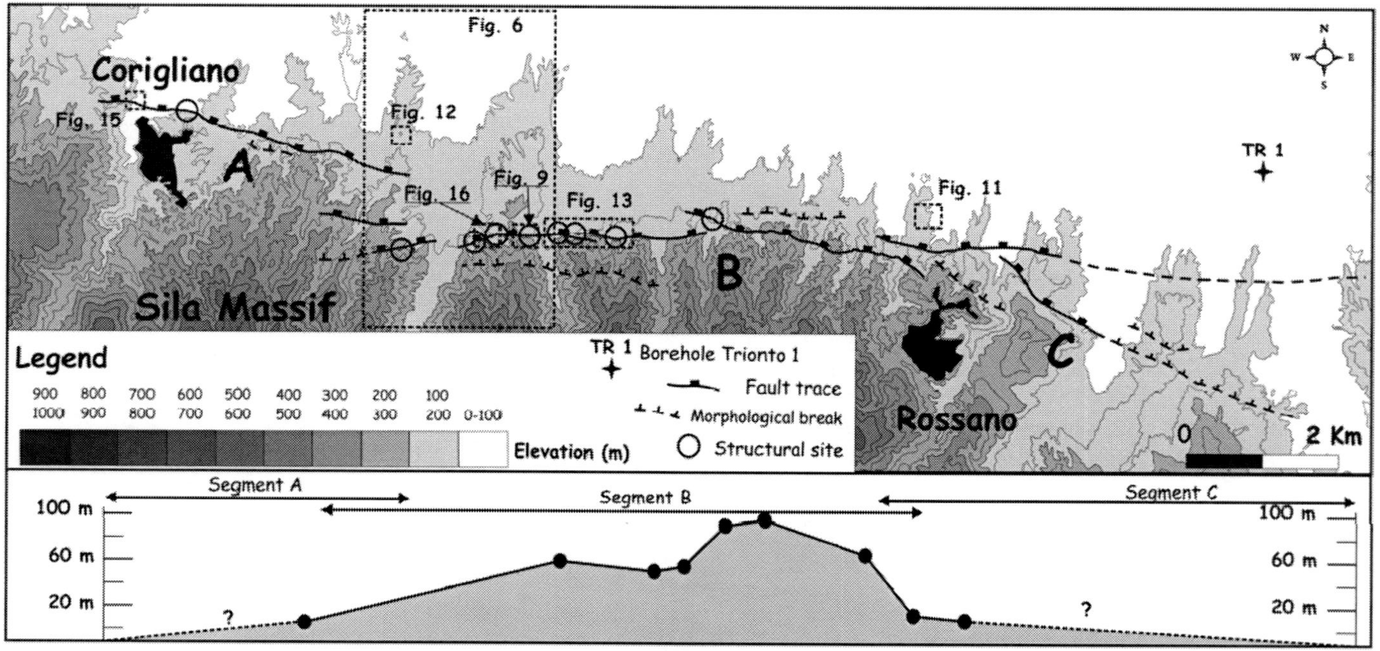

Figure 5. Rossano Fault segments (a=Corigliano segment, b=Rossano segment, c=Crosia segment) and locations of the structural sites and figures. TR-1 is the location of the deep borehole Trionto 1, reaching the crystalline basement at ~1600 m below s.l. In the figure is also shown the distribution along fault of the scarp height.

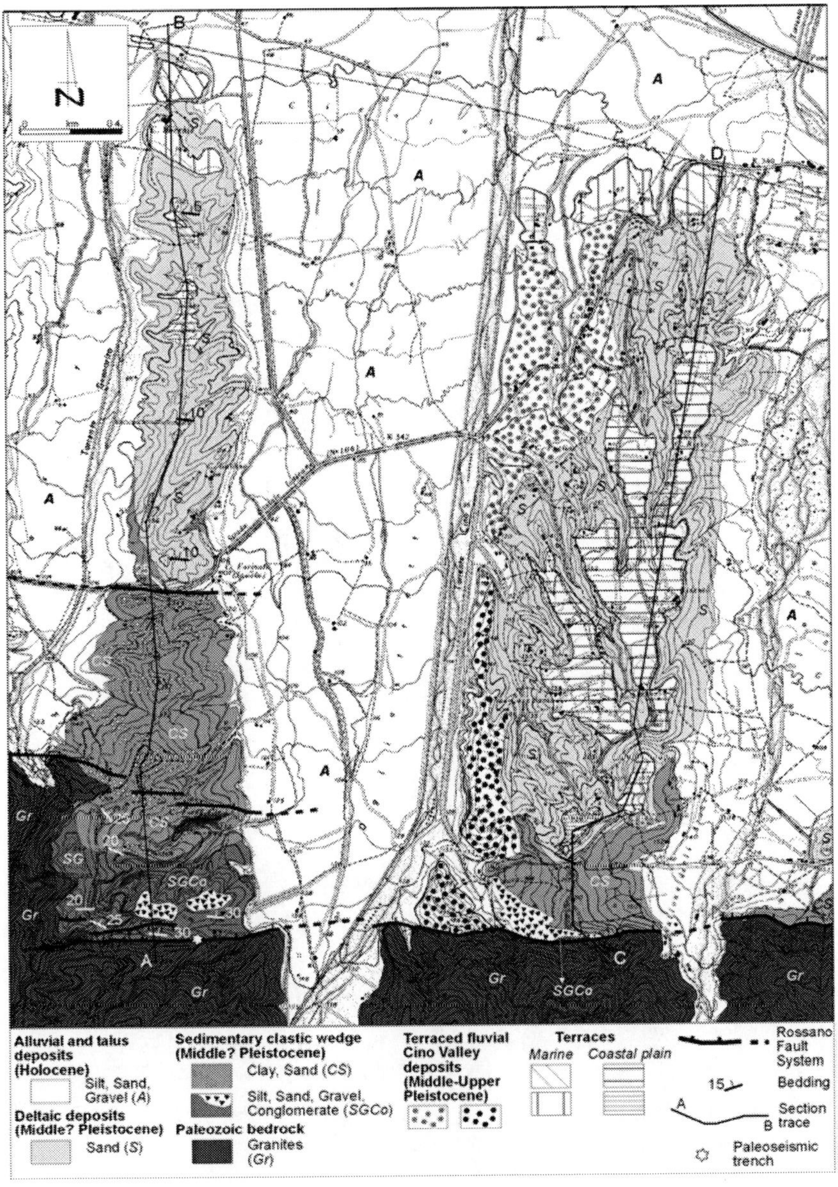

Figure 6. Geological sketch of the relay zone between the Corigliano and Rossano segments (location in figure 5), where the Quaternary marine and fluvial terraces are extensively exposed. The trace of sections of figures 7-8 are also shown.

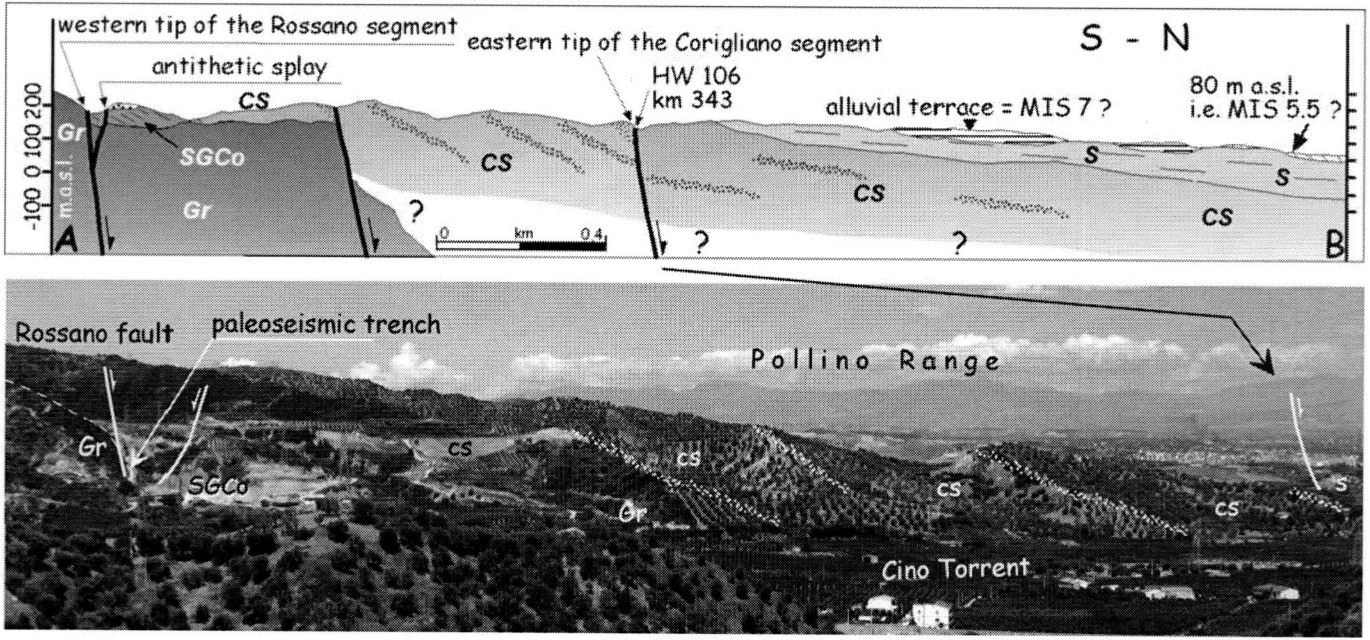

Figure 7. Simplified geological cross section across the Rossano Fault (A-B in figure 6; legend same as figure 6). The section shows the overlap zone between the Corigliano and Rossano segments. Note the Middle(?)-Pleistocene marine succession, dipping seaward, faulted against the crystalline Sila basement. The possible inner-edge of the marine terrace MIS 5.5 is showed. The lower panel is a picture of the left sector of the geological sketch.

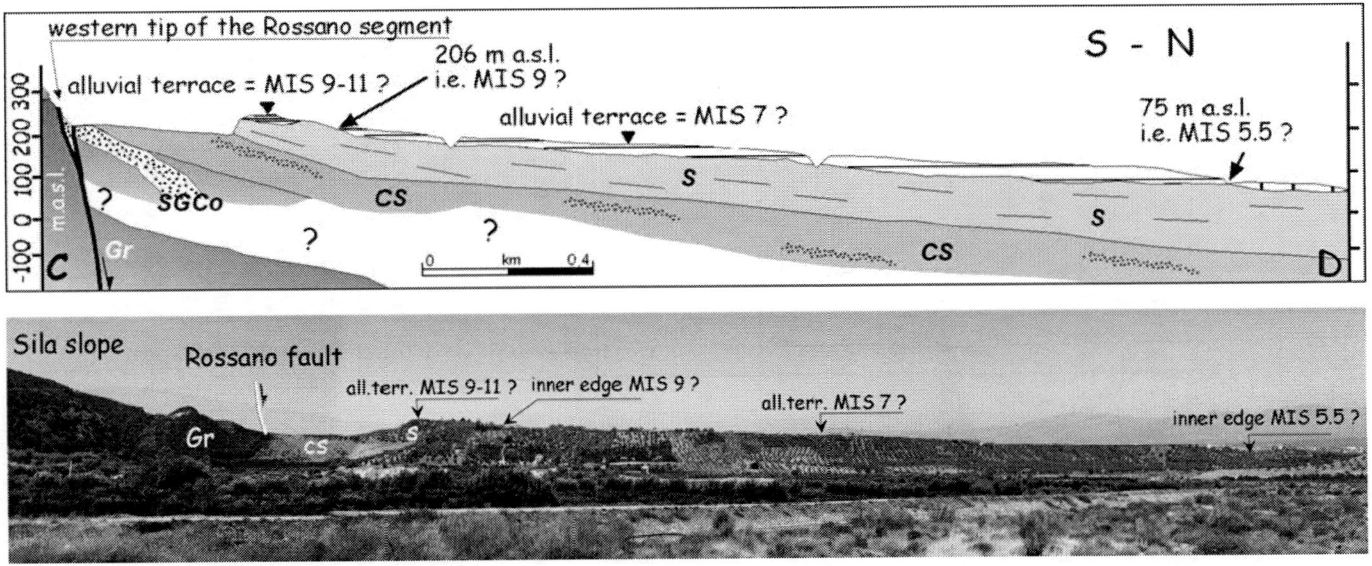

Figure 8. Simplified geological cross section across the Rossano Fault (C-D in figure 6). Note in this section the flat paleosurface topping the Pleistocene sands (S), tentatively interpreted as an alluvial coastal plain terrace, developed over a previous marine erosional terrace. The possible inner-edge of this marine terrace (MIS 9) is showed, together with the one of MIS 5.5. Lower panel is a picture of the geological sketch.

Figure 9. View looking east of the western tip of the Rossano Fault zone (location in figure 5). The Middle Pleistocene fan-delta succession (conglomerates and sandy silts and gravels, partly laying over the granite basement; right side) is displaced by several ~E-W normal faults. These faults downthrow definitely the Sila Massif below the coastal plain (left side), where the crystalline bedrock has been reached in borehole at 1600 m below s.l.

3.1. The Terrace Sequences of the Rossano Coast (Marine *vs* Alluvial?)

An impressive flight of marine and alluvial terraces characterizes the coastal sector of the investigated area. Terraces formed and were preserved because of the strong regional uplift affecting this sector of Calabria since the Middle Pleistocene, and particularly after 0.4 Ma (Carobene, 2003). Their correct identification and dating would represent a key point for the evaluation of differential crustal vertical movements (i.e. in the hangingwall of the RF).

Actually, these paleosurfaces are quite complex, and no agreement exists amongst authors about their age and nature. According to Vezzani (1968; who identifies seven orders of surfaces), and Cucci (2004; four orders), the terraces paralleling the coastline between Corigliano and Rossano are marine, as are those investigated and described in detail by Carobene (2003) farther east (four order sequence along the right side of the Trionto River). Both Cucci (2004) and Carobene (2003), by correlating the age of the interglacial peaks with the variation of the paleo-climate curves, inferred the age of each single marine terraces. However, in the overlap area of these studies, elevation (or age) of the inner edges does not fit each other. For instance, the elevation of the presumed MIS 5.5 highstand (~125 ka) is ~70 m a.s.l. in Carobene (2003), but is ~130 m a.s.l. in Cucci (2004). Conversely, following Molin et al., (2002), Corbi et al., (2008), and Robustelli et al., (2009), on the basis of the nature of deposits in which they are carved, these paleosurfaces are mainly alluvial terraces. These authors also provide ages and elevation of fluvial terraces that are roughly consistent with the corresponding marine highstands evaluated by Carobene (2003). Considering the complex and controversial state of the art, we performed an independent aerial photo study of the terraced surfaces, followed by field survey, strictly focused toward the understanding of the possible interactions between the RF and the existing flight of terraces. Our data are summarized in figure 10, where seventeen topographical sections show the existence of several "scarps" parallelling the present shoreline in the Rossano area, all interrupting the flat profile of the paleosurfaces. The elevation reported on the map (meters a.s.l.) refers to the bottom of these scarps (arrows) or to the top of the uphill truncated surface (+ symbol; i.e., in case of missing uphill scarp). Terraces and scarps are mainly carved in the remnant divides of the streams, that is in the Middle Pleistocene (?) gravel-clay-sand succession (upper part of the Ciclo Suprapliocenico-Pleistocenico in Vezzani, 1968; part of the clayey formation in Carobene, 2003; "sedimentary clastic wedge" in Robustelli et al., 2009), and end uphill against the Rossano fault scarp (actually, they end before reaching the granite hillside, because of the selective and accelerated erosion between the crystalline bedrock of the footwall and the Quaternary succession in the hangingwall). The bottom of some of these scarps occurs at comparable elevations in different transects, whereas others scarps appear only in some profiles (figure 10). We checked some of these cases in the field, finding two types of situation: 1) some of these scarps fit with secondary faults affecting both the marine substratum and the alluvial-marine terraced deposits (e.g., north of Rossano, or right bank of both Colognati and Coserie torrents; figure 11); 2) others scarps have their bottom that matches with buried upward-coarsening and tapering, reddish gravel deposits (with imbricated, flattened boulders, with lithodomes holes in the carbonate clast), sealed by finer colluvial wedge, which we though to represent – in the whole - paleoshore *facies* (e.g. in figure 12). Therefore, we interpret the former as retreated fault-scarps, while the latter could be inner-edges of ancient marine terraces. As far as the nature of terraces is concerned, in our opinion, most of the marine abrasive surfaces are now obliterated by fluvial morphologies and deposits. Therefore, part of the terraces of the investigated area are remnants of alluvial coastal-plain surfaces, with fluvial gravel and sand deposits (up to several meters thick) overlying and burying older, raised marine terraces (see also Molin et al., 2002).

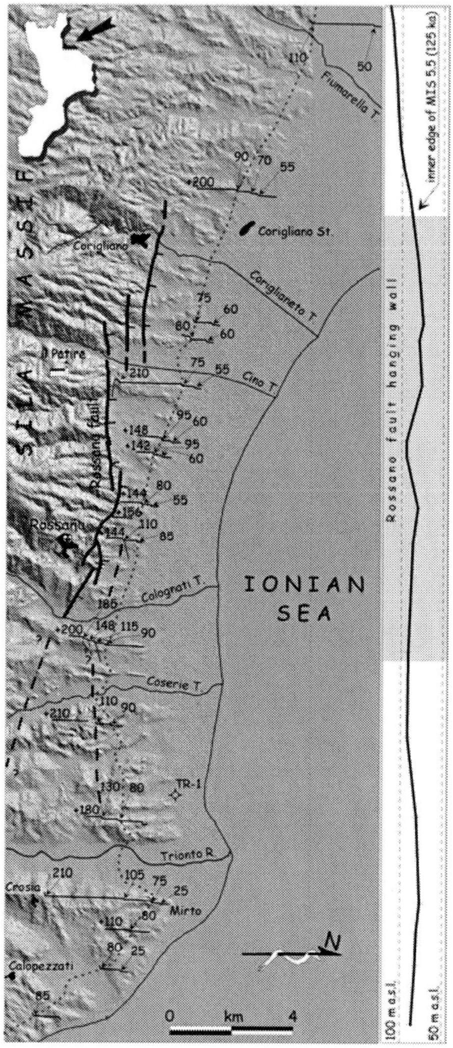

Figure 10. Shaded relief image of the Rossano coast. The main segments of the Rossano Fault are shown, together with seventeen topographical sketches drawn across the flight of marine/alluvional terraces of the area. Elevations in meters a.s.l. refer to the bottom of each identified scarp (arrows) that – in case of marine terrace – should indicate the inner edge of the relative MIS. Certain fault-scarps are evidenced on the sections by a line dipping seaward (dashed where inferred). Elevations marked by the symbol + indicate that the surface has not a scarp uphill. The right panel shows the projected profile of the presumed MIS 5.5 coastline, which appear lowered by the Rossano Fault with respect to the western (top) area.

Figure 11. View looking west of one of the faults affecting the flight of marine/alluvial terrace close to the Rossano town (location in figure 5). Note the clasts of the terrace deposits dragged along the fault zone.

Figure 12. View looking west of a fresh outcrop (location in figure 5) at the top of the terrace reasonably belonging to MIS 5.5 (see Carobene, 2003), at elevation ~80 m a.s.l. Note the gravel body pinching out uphill, marking the possible inner edge of the marine highstand. Boulders and cobbles are interested by lithodomes holes.

Thus, the inner edge of the marine terraces that are coeval to these alluvial terraces should stay below the outer edge of the alluvial surfaces (i.e., at the bottom of the scarps shown in figure 10). However, fluvial erosion and deposition have erased and/or buried most of the inner edge traces. This could explain why we did not recognise the traces of the MIS 7 paleoshoreline (quoted by Carobene, 2003 at 105-120 m), which is probably obliterated by the long and flat alluvial terraces extending up to ~200 m a.s.l. in the studied area (it is worth noting that some surfaces have an uphill termination - i.e., the bottom of the above scarp - at 110-115 m a.s.l., but these cases always fit with the presence of faults displacing the terrace deposits; see figures 10 and 11).

On the other hand, we found the buried traces of the inner edge of a marine terraces, quoted MIS 5.5 by Carobene (2003), at elevation ~80 m a.s.l. (Cino Torrent area). We correlated this outcrop to an inner edge surveyed between 95-75 m a.s.l. in the rest of the area (figure 10; note that the absolute elevation is affected by errors due to the presence of talus colluvial wedge at the base of the scarp) and to the edge reported by Cucci (2004) westward of Fiumarella Torrent at 110 m a.s.l. A lower edge, observed at 55-60 m a.s.l. between the Corigliameto and Colognati torrents, could be ascribed to a later marine highstand (i.e., MIS 5.3). Finally, in two cases, we found faint traces of an older and higher marine inner edge (marked by upward-tapering gravels, with lithodomes holes), at elevation of ~205 m a.s.l. (right bank of Cino and Colognati torrents), which could fit the MIS 9 terrace of Carobene (2003; Bigazzi and Carobene, 2004; 210 m a.s.l.). It is worth noting that other sparse flights of remnant surfaces are visible on the northern granite slopes of the Sila Massif (see also Molin et al., 2002); some of these are covered by continental gravels (sometime with boulders >1 m) and sands, and they are clearly hanging on the footwall of the RF. They could be considered relics of the apical portion of the deltaic complexes associated with the alluvial terraces of the plain, offset by the RF, and isolated by the strong erosion of the uplifted block.

4. Indications of Recent Activity of the Rossano Fault

As a result of the previous surveys, we felt more confident as far as the Late Quaternary geological evolution of the area, and then we focused our efforts along the RF. In facts, the RF has not been studied before, nor does it appear in the official geological map of the area (Burton, 1971). Only Vezzani (1968) traced a short segment fitting with the scarp showed in figure 13, whereas Molin et al. (2002) report other different strands.

We followed the surficial expression of the RF along ~12 km between the villages of Corigliano and Rossano, where the fault downthrows abruptly the northern side of the Sila massif, faulting the crystalline bedrock against the northward dipping Pleistocene marine succession (figures 2 and 8). The fault is arranged into several ~E-W segments, which slightly overlap with *en-échelon* geometry.

As mentioned before, flights of remnant paleosurfaces, together with continental and marine deposits, have been raised by the fault, and now lie hanging over the footwall (figure 13).

Although the slopes of Sila are entirely lushly forested and/or masked by agricultural works (e.g., agricultural terracing), we recognised indication of recent fault activity (e.g rock fault scarp and displaced deposits), following also the evidence gathered through the interpretation of 1954-1992's air photos. Moreover, structural data collected along the fault plane (figure 5) show two main trends, oriented N55°E and N83°E, respectively, mirroring the presence of short fault jogs (see rose diagram in figure 14), whereas *striae* orientation indicate that RF has an almost pure normal dip slip kinematics, with a direction of extension (direction of the minimum stress axis σ3) sub-horizontal and oriented N191° (figure 14).

Figure 13. View looking ESE of the Rossano Fault scarp (big arrows; location in figure 5). Rossano is the village (left-background) where historical sources described surface breaks during the 1836 earthquake. Note the prominent triangular facets which characterize this sector of the northern Sila hillside; they are probably due to combined erosive exhumation (regional uplift) and cumulated surface faulting processes. Little arrows indicate a gently dipping paleosurface carved in the granite of the RF footwall, which hangs now hundred meters over the plain.

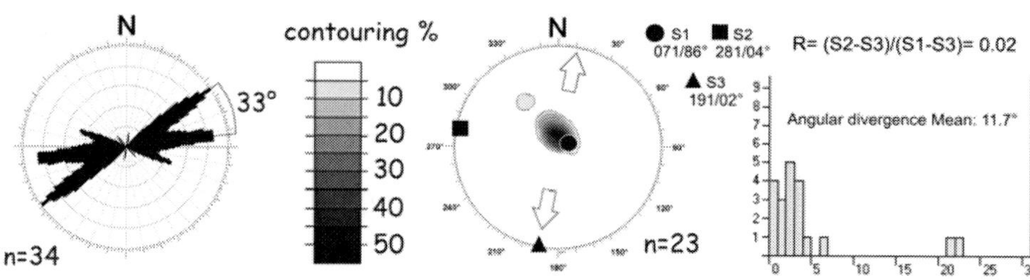

Figure 14. Synthesis of the structural analyses carried out along the Rossano Fault (see Fig. 5 for location of surveying). Dispersion of plane strike is 33° (left panel), whereas the *striae* indicate a horizontal, 191° trending σ3 axis (central panel), with a mean angular divergence of 17° (right panel).

From the morphological point of view, the fault generates an impressive rectilinear hillside, associated with prominent triangular facets carved in the crystalline bedrock (figure 13). The scarp varies between 10 m, in the tip zones of the fault, and almost 100 m, in the central part, (figure 5) with a pattern similar to the distribution of displacement along large basin bounding normal faults (Roberts, 2007). However, the genesis and height of the triangular facets must be interpreted as due to the combined action of granite-bedrock exhumation (i.e., due to the strong regional uplift) and to the cumulated coseismic surface rupture events.

A robust indication of recent activity is the presence of a basal fault scarp (*sensu* Gilbert,1884), which is equally carved in sandy-clayey Middle Pleistocene deposits (e.g., Corigliano area; see figure 15), in highly degraded granites (see figure 16), and in coarse slope deposits (see figure 17).

This scarp is usually ~2-m-high, being somewhere characterized by at least three 0.5/0.6m-high horizontal bands, each one with both different dips (from ~85° at the bottom, to ~55° at the top) and degradational aspect. Figure 16 summarizes the observations carried out along twelve sections at the base of the central portion portrayed in figure 13, and provides the average height and dip ranges of the three bands that characterize the fault free-face. A possible explanation for these features is that each band might be partly associated with a single faulting episode, so that the higher, less steep and more degraded the band, the older is the event which it is associated to (see Giaccio et al., 2002).

Figure 15. Morphological evidence of the recent fault re-activitation (location in Fig. 5). A) view looking south of the fault scarp affecting the Middle Pleistocene sandy-clayey succession near Corigliano. Considering the high erodibility of these terrains, the freshness of the scarp suggests a recent activity of the fault. B) fault-plane (arrows) along the same scarp, between the marine sands (footwall) and talus and marine clays in the hangingwall.

According to this interpretation, the free-face records at least three surface faulting events, each one with offset of ~0.5÷0.6 m. However, we can not exclude that part of this value was caused by climatogenic processes, as exhumation due to soil instability along the slope. Therefore, the measured bands represent a maximum amount of offset per event.

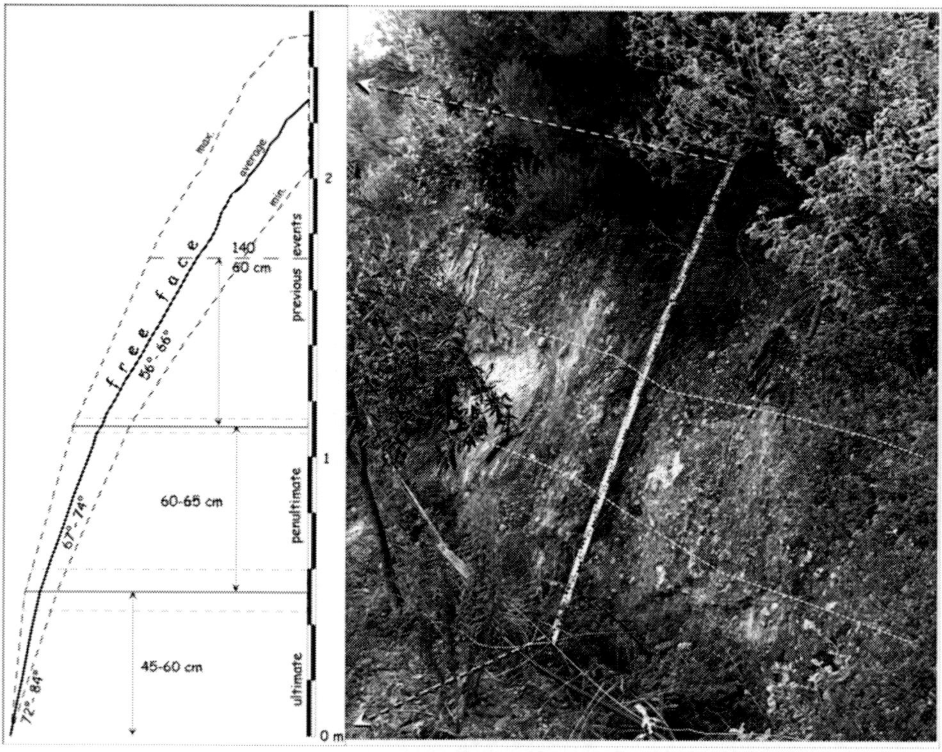

Figure 16. Free-face cropping out at the base of the Rossano Fault scarp (location in Fig. 5), and, on the left, sketch summarizing the geometrical properties of the plane (dips and heights have been averaged on twelve spaced sections). We recognized at least three ~0.6-m-high bands with different dip and degradational aspect, each one partly associable to single coseismic rupture. The upper (and oldest) one is deeply degradated, and it presents the lowest dip; thus, it might record several past surface ruptures, beside long-term erosive exhumation.

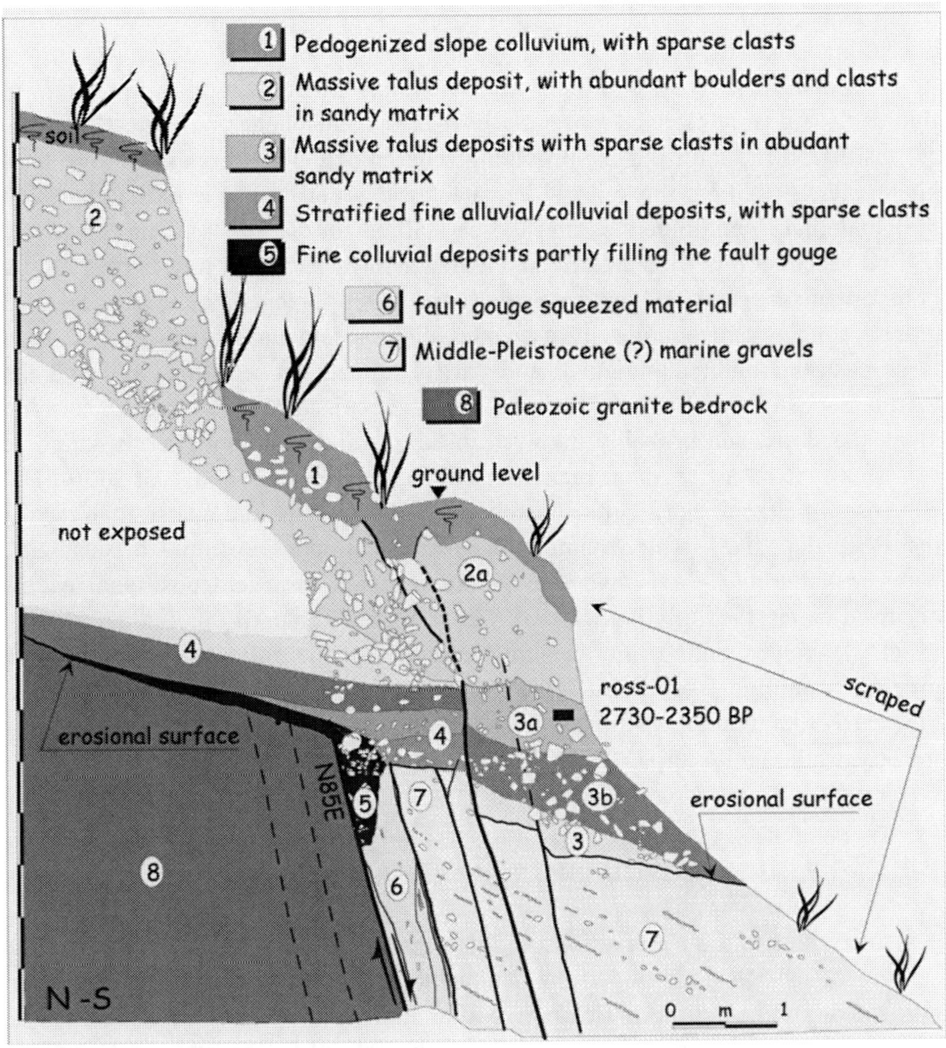

Figure 17. Sketch of the paleoseismic trench excavated across the Rossano Fault (location in Fig. 6). Even if no certain event-horizons can be recognized within the talus succession, the absolute age of a charcoal sampled in unit 3 provides the *terminus post-quem* for the latest fault ruptures.

4.1. Paleoseismic Trench

In order to ascertain the recent activity of the RF, we dug an explorative trench, exploiting the erosive bank of a natural gully at the toe of the fault-scarp

(Galli et al., 2006), finding both marine and slope deposits faulted against the crystalline rocks (figure 17).

The marine succession which we have found in the hangingwall (unit 7 in figure 17) is deeply affected by regularly spaced shear-planes; pure dip-slip *striae* are visible on the fault planes, providing reliable evidence to constraint the fault kinematics. A ~20 cm thick, argillitic cataclastic zone (unit 6) is packed between the Quaternary conglomerates and the crystalline rocks (unit 8). The top of the marine succession is truncated by an erosional surface, that in turn is overlain by both organized (unit 4) and massive (unit 3) slope-deposits, with occasional granite cobbles and boulders in a coarse, slightly cemented sandy matrix. Also these units are faulted, dragged, and partly trapped between the fault and the marine deposits. Finally, unit 2 is a faintly pedogenized rudite, currently forming along the slope, and probably involved in the last deformation. An absolute date of a small charcoal fragment picked in unit 3a provides a ^{14}C age of 2730-2350 BP (ross-01, 2σ cal. age). Unfortunately due to the texture and coarse lithology of the slope deposits - which hamper the identification of eventual offsets (e.g., between subunits 2-2a) - and due to the repeated erosive/depositional events occurred within the gully *thalweg*, we were not able to identify individual faulting events. However, taking into account the dating of the collected sample (which gives the minimum age of unit 3), the last rupturing paleoearthquakes occurred in the past 2-3 ka, being possibly associated with part of the free-face running along the base of the fault scarp.

5. Discussion and Conclusions

According to the observations and data herein presented, it emerges that the activity of the RF conditioned the sedimentary and morphological evolution of this sector of the northern Calabria during the Middle-Upper Pleistocene, i.e., after the inception of the strong ~1mm/yr regional uplift (Bigazzi and Carobene, 2004). In fact, with the exception of the Rossano area – where Tortonian deposits outcrop in the Rossano Fault footwall - the Pleistocene marine-deltaic succession is faulted directly against the crystalline basement, which – in turn - is lowered 1600 m below s.l. in the hangingwall (data from onshore borehole Trionto 1). In one of the main fault step zones, west of Cino Torrent (i.e. in the hangingwall of Rossano segment), the Middle Pleistocene sandy clays deposited directly on a paleosurface carved in the granite rocks (figure 7); here, the entire sedimentary succession hangs in the footwall of the Corigliano segment, so that the uppermost terraces (i.e., older than MIS 5.5) have been dissected by accelerated erosive processes. The lower terraces, instead, appear again in the hangingwall of the Corigliano segment, with elevations comparable to those observable eastward (figure 10). As far as the marine/alluvial terraces are concerned, in the entire hangingwall of the RF, the presumed (~buried) inner edge of MIS 5.5 is at 75-95 m a.s.l. (it is ~80 m a.s.l. in the certain outcrop of 10), whereas it is 110 m a.s.l west of Fiumarella Torrent (figure 10).

Moreover, according to Cucci (2004), farther west, in the Sibari Straits the inner edge of MIS 5.5 highstand crops out at 130-140 m a.s.l., that is ~50-60 m higher than in the Rossano area. Considering the short distance of the MIS 5.5 paleoshoreline from the RF, if this elevation difference could be tentatively ascribed to the hangingwall downthrown, the rough slip-rate during Late Pleistocene would be ~0.5 mm/yr. On the other hand, considering that the elevations of the presumed MIS 5.5 remain the same proceeding toward Mirto (Carobene, 2003), it is possible to hypothesize that an eastern "hidden" step of the RF might exist south of Calopezzati and Crosia villages (c in figure 5; dashed line in figure 10), where only faint scarps were recognised. In this area, the surficial expression of this structure could be masked by erosion and landsliding processes in the sandy-clayey succession (conversely, it is magnified by the abrupt contact between hard crystalline and soft Pleistocene deposits along the RF). However, no certain field indications exist concerning this issue, which will need future investigation.

As far the present activity of the RF is concerned, morphological (i.e., rectilinear fault-scarp, triangular facets, fault free-face), and geological data (late Pleistocene faulted deposits) suggest recent surface ruptures, at least between the villages of Corigliano and Rossano (for a length of ~12 km). Actually, the explorative paleoseismic trench that we dug across the western sector of the fault confirmed the rupture during late Holocene time, at least after ~3 ka BP. In particular, systematic observations along the rock fault scarp showed steady dip-variations from the bottom to the top of the free-face, each one associated with different degradational aspect of the plane; we tentatively associate part of these features with as many consecutive surface rupture.

At the same time, documents research in archives and libraries, coupled with the reinterpretation of all the available primary historical sources, allowed the reappraisal of the effects of the strongest known earthquake of the area (April 24, 1836). The 1836 HIDD matches well with the RF trend, suggesting this normal fault to be the causative source for the earthquake. Moreover, coeval accounts on sea-waves, liquefactions and surficial breaks have been compared with the RF location and kinematics, providing further indication concerning the coseismic rupture of this structure in the 1836 event. Fault length and average co-seismic offset allow us to calculate a rough value of magnitude associated to this structure. According to the empirical relationships of Wells and Coppersmith (1994), we obtain Mw=6.28, which is close to the value expected using the regression law calculated by Galli et al (2008) for "Apennine faults" (Mw=6.13). Both values are consistent with the macroseismic equivalent magnitude evaluated by using the Boxer algorithm (Gasperini, 2002), applied to our HIDD (Mw=6.24). Based on our seismotectonic interpretation, the epicentre of the 1836 earthquake (currently located near the village of Crosia) would fall farther west, in the hangingwall of the RF (namely at average coordinates 16.58°E 39.62°N); the strong effects in the small villages of Crosia and Calopezzati might be explained with directivity effects, i.e., with an eastward rupture (from Corigliano – that suffered a relatively low intensity - to Rossano; see arrow in figure 2). The current activity of the RF is strengthened also by the existence of the 1995 event at the eastern tip of the fault, the focal mechanism of which suggests the rupture of an E-W oriented fault plane with extensional kinematics (i.e., like the RF; see figure 2).

Finally, considering the number of bands observed along the ~2-m-high free-face, the reasonably recent age of the fresh rock fault scarp (which is conserved both in clayey and coarse deposits, and in granite rocks), and the occurrence of historical earthquakes in the area, it is possible to hypothesize that at least three events occurred in the past ~3 ka, the last ones being the ~951 and 1836 earthquakes, respectively. This gives a return time of ~0.9 ka for Mw=6.2 events on the RF, which, in turn, would be characterized by ~0.6 mm/yr of slip-rate.

As a concluding remark, basing on both kinematics of the RF and fault slip data coming from other active faults (i.e. Lakes Fault in figures 1 and 4), and considering also the contribution of the WNW-ESE Pollino normal fault (figure 1), we observe that the central and eastern sector of northern Calabria is currently experiencing ~N-S extension, in agreement with the different GPS velocities calculated by D'Agostino and Selvaggi (2004) for central-southern Calabria (3.2-3.3 mm/yr in a NNW direction) and Apulia (4.6-4.8 mm/yr in a NNW direction).

As widely demonstrated in several others published works concerning earthquake geology, we think that a multidisciplinary approach - such that consisting of archive research (historical seismology), structural and geomorphological analyses, and paleoseismology - is a robust tool to identify, and to characterize seismogenic sources, providing also useful data for seismic hazard assessment. However, it is our opinion that additional paleoseismological investigation across the RF are needed to better constrain fault parameters (i.e. recurrence time, fault slip rate) and to reduce the uncertainty in terms of the recent seismogenic behaviour of this structure.

ACKNOWLEDGMENTS

We are grateful to F. Dramis, G. Robustelli and G. Fubelli for the useful discussion and suggestions. Thanks also to R. Crossley for the English review, and to G. Mastronuzzi for providing us library documents.

REFERENCES

Amodio Morelli, L., Bonardi, G., Colonna, V., Dietrich, D., Giunta, G., Ippolito, F., Liguori, V., Lorenzoni, S., Paglionico, A., Perrone, V., Piccarreta, G., Russo, M., Scandone, P., Zanettin Lorenzoni, E., & Zuppetta, A. (1976). L'arco Calabro-Peloritano nell'orogene appenninico Maghrebide. *Mem. Soc. Geol. It., 17*, 1-60.

Argnani, A., Brancolini, G., Rovere, M., Accaino, F., Zgur, F., Grossi, M., Fanzutti, F., Visnovic, P., Sorgo, D., Lodolo, E., Bonazzi, C. & Mitchell, N. (2008). Hints on active tectonics in the Messina Straits and surroundings: preliminary results from the TAORMINA- 2006 seismic cruise. *Boll. Geof. Teorica Appl., 49*, 163-176.

Bartholomaeus Rossanensis, (11th cent). *του οσιου πατροσ εμον Νειλου του νεου: codice greco criptense B. B. 2. di S. Bartolomeo*, G. Giovanelli (ed.), Badia di Grottaferrata, Roma 1972.

Bigazzi, G. & Carobene, L. (2004). Datazione di un livello cineritico del Pleistocene medio: relazioni con sedimentazione, sollevamento e terrazzi marini nell'area Crosia-Calopezzati in Calabria (Italia). *Il Quaternario, 17*, 151-163.

Burton, A.N. (1971). Carta geologica della Calabria alla scala 1:25.000. Relazione Generale. Cassa per il Mezzogiorno, Servizio Bonifiche, Roma.

Carobene, L. (2003). Genesi, età, sollevamento ed erosione dei terrazzi marini di Crosia-Calopezzati (Costa Ionica della Calabria - Italia). *Il Quaternario, 16*, 43 -90.

Corbi, F., Fubelli, G., Lucà, F., Robustelli, G., De Martini, P.M., Muto, F., Scarciglia, F. & Dramis, F. (2008). Tectonic and climate control on quaternary evolution of the north-eastern Sila Massif (ionian Calabria, Italy). EGU General Assembly 2008, *Geophysical Research Abstracts, 10*, EGU2008-A-07098.

Cucci, L. (2004). Raised marine terraces in the Northern Calabrian Arc (Southern Italy): a ~ 600 kyr-long geological record of regional uplift. *Annals of Geophysics, 47*, 1391-1406.

D'Agostino, N. & Selvaggi, G. (2004). Crustal motion along the Eurasia-Nubia plate boundary in the Calabrian Arc and Sicily and active extension in the Messina Straits from GPS measurements, *Journal. Geophys. Res., 109*, B11402, doi:10.10.1029/2004JB002998.

De Rosis, L. (1838). Descrizione del tremuoto avvenuto in aprile del 1836, In: *Cenno storico della città di Rossano e delle sue nobili famiglie, 1*, pp.88-112, Napoli.

Dewey, J.F., Helman, M.L., Turco, E., Hutton, D.H.W., & Knott, S.D. (1989). Kinematics of the western Mediterranean, , In M.P. Coward, D. Dietrich and R.G. Park (Eds.), *Alpine tectonics*, Geol. Soc. Spec. Publ., London, 45, 265-283.

Galli, P. (2000). New empirical relationships between magnitude and distance for liquefaction. *Tectonophysics, 324*, 169-187.

Galli, P. & Bosi, V (2002). Paleoseismology along the Cittanova fault. Implications for seismotectonics and earthquake recurrence in Calabria (southern Italy). *Journ. of Geophys. Res., 107*, B3, doi: 10.1029/2001JB000234.

Galli, P. & Bosi, V. (2003). The catastrophic 1638 earthquakes in Calabria (southern Italy). New insight from paleoseismological investigation. *Journ. of Geophys. Res., 108*, B1, doi: 10.1029/2002JB01713.

Galli, P., Galadini, F. & Pantosti, D. (2008). Twenty years of paleoseismology in Italy, *Earth Science Review, 88*, 89-117, doi:10.1016/j.earscirev.2008.01.001.

Galli, P., Ruga, A., Scionti, V. & Spadea, R. (2006). Archaeoseismic evidence for a Late Roman earthquake in the Crotone area (Ionian Calabria, southern Italy): seismotectonic implication. *Journ. of Seismology, 10*, 443-458.

Galli, P., Scionti, V. & Spina, V. (2007). New paleoseismic data from the Lakes and Serre faults (Calabria, southern Italy). Seismotectonic implication. *Boll. Soc. Geol. It., 126*, 347-364.

Gasperini, P. (2002). The Boxer program, release 3.02 available at: http://ibogfs.df.unibo.it/user2 /paolo/www/boxer/boxer.html.

Ghisetti, F. (1981). Upper Pliocene–Pleistocene uplift rates as indicators of neotectonic pattern: an example from Southern Calabria (Italy). *Zeithschrift fur Geomorphologie 40*, 93–118.

Giaccio, B., Galadini, F., Sposato, A., Messina, P., Moro, M., Zreda, M., Cittadini, A., Salvi, S., & Todero, A. (2002). Image processing and roughness analysis of exposed bedrock fault planes as a tool for paleoseismological analysis: results from the Campo Felice fault (central Apennines, Italy). *Geomorphology, 49*, 281-301.

Gilbert, G.K. (1984). A theory of the earthquakes of the Great Basin, with practical application, *American Journal of Science, 27*, 157.

Jolivet, L. & Faccenna, C. (2000). Mediterranean extension and the Africa-Eurasia collision. *Tectonics, 19*, 1095-1106.

Messina, A, Russo, S., Perrone, V., & Giacobbe, A. (1991). Calc-alkaline Late Variscan two micacordierite-Al silicate-bearing intrusion of the Sila batholith (northern sector of the Calabrian Arc, Italy). *Boll. Soc. Geol. It., 110*, 365-389.

Molin, P., Dramis, F., & Lupia Palmieri, E. (2002). The Pliocene-Quaternary uplift of the Ionian northern Calabria coastal belt between Corigliano Calabro and Capo Trionto. *Studi geologici Camerti, 2/2*, 135-145.

Nakamura, Y. (1989). A Method for Dynamic Characteristics Estimation of Subsurface using Microtremor on the Ground Surface, Quarterly Report of Railway Technical Research Institute (RTRI), 30, #1.

Roberts, G. P. (2007). Fault orientation variations along the strike of active normal fault systems in Italy and Greece: Implications for predicting the orientations of subseismic-resolution faults in hydrocarbon reservoirs. *American Association of Petroleum Geologists Bulletin, 91*, N. 1, 1-20.

Robustelli, G., Lucà, F., Corbi, F., Pelle, T., Dramis F., Scarciglia, F., Fubelli, G., Muto, F. & Cugliari, D. (2009). Alluvial terraces on the Ionian coast of northern Calabria, southern Italy: Implications for tectonic and sea level controls. *Geomorphology, doi:10.1016/j.geomorph.2008.12.010, in press.*

Roda, C. (1964). Distribuzione e facies dei sedimenti neogenici nel bacino crotonese. *Geologica Romana, 3*, 319-366.

Romano, M. (1836). Al S.r Sotto Intendente, Rossano, 10 Luglio 1836. Archivio Comunale di Rossano, Registro delle Deliberazioni Decurionali degli anni 1834, 1835 e 1836, ff. 159-161.

Rossetti, F., Faccenna, C., Goffè, B., Monié, P., Argentieri, A., Funiciello, & R. Mattei, M. (2001). Alpine structural and metamorphic segnature of the Sila Piccola Massif nappe stack (Calabria, Italy): Insights for the tectonic evolution of the Calabrian Arc. *Tectonics, 20(1)*, 112-133.

Rossi, A.A. (1836). De' tremuoti nella Calabria Citeriore l'anno 1836. In: Annali Civili del Regno delle Due Sicilie, 12(23), Napoli, pp.12-34.

Scionti, V., Galli, P. & Chiodo, G. (2006). The Calabrian seismicity during the Viceroyalty of Naples: sources silence or silent sources? The case of the strong 1744 earthquake. *Boll. Geof. Teor. e Appl., 47*, 53-72.

Stucchi, M. & Albini, P. (2000). Quanti terremoti abbiamo perso nell'ultimo millennio? In Galadini, Meletti & Rebez (Ed.), *Le ricerche del GNDT nel campo della pericolosità sismica (1996-1999),* CNR-GNDT, 333-343.

Tortorici, L., Monaco, C., Tansi & Cocina, O. (1995). Recent and active tectonis in the Calabrian arc (Southern Italy). *Tectonophysics, 234*, 37-55.

Van Dijk, J. & Okkes, M. (1991). Neogene tectonostratigraphy and kinematics of Calabrian basins; implications for the geodynamics of the Central Mediterranean. *Tectonophysics 196*, 23-60.

Van Dijk , J.P., Bello, M., Brancaleoni, G.P., Cantarella, G., Costa, V., Frixa, A., Golfetto, F., Merlini, S., Riva, M., Torricelli, S., Toscano, C. & Zerilli A. (2000). A regional structural model for the northern sector of the Calabrian Arc (southern Italy). *Tectonophysics, 324*, 267-320.

Vannucci, G. & Gasperini, P. (2004). The new release of the database of Earthquake Mechanisms of the Mediterranean Area (EMMA Version 2), *Annals of Geophysics, 47*, 307-334.

Vezzani, L. (1968). I terreni plio-pleistocenici del basso Crati (Cosenza). *Atti Acc. Gioenia Sc. Nat., VI*, 28-84.

Wells, D. & Coppersmith, K. (1994). New empirical relationships among magnitude, rupture length, rupture width, rupture area and surface displacement. *Bull. Seismol. Soc. Am., 84*, 974-1002.

Working Group CPTI (2004). Catalogo Parametrico dei Terremoti Italiani (versione 2004, CPTI04): Internet website INGV Sezione di Milano, http://emidius.mi.ingv.it/CPTI/ (last revision 27 May 2004)

In: Recent Progress on Earthquake Geology
Editor: Pierpaolo Guarnieri, pp. 79-108
ISBN: 978-1-60876-147-0

Chapter 4

SLIP RATE AND EARTHQUAKE RECURRENCE OF THE DÜZCE FAULT (NORTH ANATOLIAN FAULT ZONE): INTEGRATING GEOMORPHOLOGICAL AND PALEOSEISMOLOGICAL ANALYSES

S. Pucci**, D. Pantosti and P.M. De Martini
Istituto Nazionale di Geofisica e Vulcanologia,
Via di Vigna Murata 605, 00143 Rome, Italy

ABSTRACT

To learn about recurrence of large earthquakes and strain model on the Düzce segment of the North Anatolian fault zone, which ruptured on November 12, 1999 (Mw 7.1), systematic geomorphological and paleoseismological analyses were integrated.

In order to provide new estimates on Quaternary slip rate of the fault, geomorphological mapping along the fault trace permitted analysis of fault-related cumulative landforms and drainage pattern settings. Remnants of an old alluvial fan modeled by fluvial terraces and 41 right-hand stream deflections were reconstructed, described and used as offset geomorphic

* Corresponding author. Current address: Istituto Nazionale di Geofisica e Vulcanologia, Via di Vigna Murata 605, 00143 Rome, Italy. Tel.: +39-6-51860406; Fax: +39-6-51860507; E-mail address: stefano.pucci@ingv.it (S. Pucci)

markers. Two correlated Late Pleistocene terrace risers, offset about 300 and 900 m, respectively, were dated by means of OSL method about 21 kyr BP and 60 kyr BP. Moreover, the onset of the offset of the streams deflected for a total of ~100 m was radiocarbon dated about 7000 yrs B.P. These data translate to a constant rate of deformation of 15.0±3.2 mm/yr for the last 60 kyr. On the basis of a "characteristic-earthquake model" and under constant slip rate assumptions, it is possible to estimate that stream deflections across the Düzce fault may be explained by repetition of 20–30 1999-like earthquakes during the last 7000 years; thus, the recurrence time for surface rupturing events of the Düzce fault is 290±60 years.

With the aim to reconstruct the record of last large earthquakes, 10 trenches at five sites were excavated. By merging information obtained from all trenches, evidence for three surface faulting earthquakes prior to 1999 were found. These paleoearthquakes are dated on the basis of radiocarbon, 210Pb and archaeological information and can be summarized as follows: 1) 1685–1900 A.D., possibly the end of the nineteenth century; 2) 1685–1900 A.D., possibly close to 1700 A.D.; 3) 800–1000 A.D. Some of them can be correlated to historical earthquakes that occurred in 967, 1719, 1878 or 1894 A.D. These paleoseismological results, merged with those from previous papers, are suggestive of bimodal recurrence distribution yielding overall average recurrence time of ~350 yrs for the past two millennia, compatible with that calculated from the geomorphic markers analysis. Under the assumption of "characteristic earthquake," bimodal paleoearthquake distributions indicate slip rate fluctuations during the past two millennia, with clustered high strain release. These results suggest the Wallace-type strain release model for the Düzce fault, and average strain accumulation of 11.4±1.2 mm/yr, in agreement with slip rate results for the past 60 kyr obtained by geomorphic marker analysis.

1. Introduction

Slip rates of active faults and earthquake recurrence are generally estimated through trenching investigations and analysis of offset geomorphic markers.

Both approaches have limitations. Although very powerful, investigation by paleoseismologic trenching contains a fundamental limit that strongly affects estimates of slip rates, recurrence intervals and behavioural models of strain release. Such limit is the usual small number of paleoearthquakes that can be recognized, to cover an insufficient length of time to be statistically trustworthy. However, it is possible to reduce uncertainties spreading the analyses to the entire fault extent and providing accurate paleoearthquake dating. Despite that, while trench investigations provide age of paleoearthquakes and possibly slip per event of dislocation cumulated during several events, observations generally cannot be extended for longer than short-term periods (<10,000 years). Conversely, offset geomorphic markers allow measurements of long time-scale strain accumulation, with no information on the timing and patterns of faulting events. Studies of such markers provide averaged information on the slip rate or the cumulative slip during repeated events, spanning analysis over middle- and long-term periods (>100,000 years). For this reason, it is critical to overcome such limitations through the integration and comparison of these independent approaches.

For this purpose, to learn about slip rates and recurrence of large earthquakes on the Düzce segment of the North Anatolian fault zone, which ruptured on November, 12, 1999 (Mw 7.1), systematic geomorphological and paleoseismological analyses were integrated. This is based on the fact that earthquakes are an integral component of the landscape evolution; this is true all along the North Anatolian fault zone (NAFZ) and it is very clear for the Düzce fault. In fact, the outstanding morphology of this active strike-slip fault is one of the direct consequences of (1) long-lasting significant deformation, (2) important present-day strain-rate, (3) frequent seismicity that ordinarily accommodates the present-day strain rate, and (4) recurrent surface faulting potentially recorded by the deformation of recent deposits.

The area of the Düzce segment was favourable for slip rate studies because of (1) extensively faulted Pleistocene and Holocene continental deposits, (2) high preservation potential of the landforms with respect to the time-scale of the tectonic process, and (3) near-fault deformation recorded and expressed by the landscape as recognized following the 1999 earthquake [Pucci et al., 2006; Pucci, 2007], which offer useful records of long-term slip rates.

In the following, after a brief introduction on the tectonic framework of the NAFZ and on the Düzce fault segment, we present the analysis of the preliminary Düzce fault slip rate and recurrence interval estimates obtained through geomorphological and paleoseismological studies. In fact, the recognition of lateral displacements of streams and alluvial terraces, resulting from the strike-slip faulting activity during the past 60 ky [Pucci et al., 2008] and the reconstruction of the record of past large earthquakes, by means of 10 trenches excavated along the entire fault length [Pantosti et al., 2008], allow such integrated study of an independent dataset.

2. Geological Setting

The NAFZ is an active right-lateral system, about 1500 km long, which bounds the westward-extruding Anatolian block to the north [McKenzie, 1972; Şengör, 1979; Barka, 1992; Şaroğlu et al., 1992] (Figure 1) that, based on the age of fault-related basin deposits, is considered to have been active since the Late Miocene [Şengör et al., 1985; Dewey et al., 1986; Şengör et al., 2005] or Early Pliocene [Barka, 1992]. Seismicity along the NAFZ is characterized by frequent moderate to large earthquakes (M 7) with focal mechanism solutions that show essentially pure right-lateral strike-slip motion [Canıtez and Uçer, 1967; McKenzie, 1972; Reinecker et al., 2004].

The trace of the NAFZ splays progressively to the west up to three major strands in the Marmara Sea (NNAF, CNAF, and SNAF in Figure 1) [Wong et al., 1995; Armijo et al., 1999; Okay et al., 1999]. However regional GPS networks indicate the northern strand to accommodate most of the 20–30 mm/yr strike-slip strain [Straub et al., 1997; Reilinger et al., 1997, 2000; Kahle et al., 1999, 2000; McClusky et al., 2000]. Similar deformation rates of 20±10 mm/yr [Jackson and McKenzie, 1988; Westaway, 1994; Parson et al., 2000] are suggested by seismic moment release over the past 100–400 yrs.

Differently, the slip rates documented from geology and geomorphology along the NAFZ are much smaller than the present-day geodetic measurements, showing a large uncertainties and variability (from 2 mm/yr up to 24 mm/yr). This difference suggests an acceleration of the deformation over time or an underestimation in measuring and dating geological offsets, as testified by a limited spatial and age distribution of the studied geologic and geomorphologic markers [see Pucci et al., 2008].

During the twentieth century, a progression of large earthquakes migrating westward occurred along the NAFZ, that culminated on the destructive 1999 Izmit seismic sequences (Figure 1). The second mainshock of this sequence, the Mw 7.1 on 12 November 1999, ruptured the Düzce fault segment. This earthquake is considered to have been triggered by the Mw 7.4 Izmit earthquake that occurred 3 months earlier (17 August) on the next fault segment to the west. The Düzce earthquake produced a ~40-km-long surface rupture with up to 5 m of right-lateral offset and up to 2.5 m of vertical throw [Akyüz et al., 2000, 2002; Pucci et al., 2007] (Figure 2a). The horizontal coseismic slip distribution curve (Figure 2b), built by integrating the data collected after the 1999 earthquake, exhibits a large along–strike variability, with 5.0 m maximum and 2.4 m dextral offsets averaged on the entire fault length. The main characteristics of the slip distribution, expressed by the interpolation of the maximum values that better describes the seismic moment at the surface (for details see Pucci et al., 2007), are: 1) a bow-shape, with slip tapering off at both ends (west of Efteni Lake and east of Kaynasli); 2) the central part, with a constant mean value of about 4.0 m.

The Düzce fault is located just west of the Bolu basin, where the NAFZ starts splaying into two main strands, the Düzce/Karadere to the north and the Mudurnu to the south, that equally take up the present-day deformation [Ayhan et al., 1999, 2001; Pucci et al., 2008]. The Düzce fault has an average E-W trend and a clear geomorphic expression, being the edge between the Quaternary Düzce and Kaynasli basins to the north and the Paleozoic-Eocene rocks of the Almacik block to the south (Figure 2a). The western and eastern boundaries of the 1999 Düzce earthquake rupture coincide to the releasing fault junctions [*sensu* Christie-Blick and Biddle, 1985] and complex step-overs, respectively, that they form with the northern strand of the NAFZ.

Although Turkey has one of the richer records of historical seismicity in the Mediterranean, no clear evidence for historical earthquakes produced by the Düzce segment of the NAFZ during the past centuries has been found. The only historical earthquakes that are known to be close enough to be potentially associated to the Düzce fault are A.D. 967, A.D. 1719, A.D. 1754, A.D. 1878, A.D. 1894 [Ambraseys and Finkel, 1995; Ambraseys, 2002, 2006] (Figure 1).

3. Geomorphological Data

3.1. The Düzce Fault Landforms

We carried out geomorphic analysis along the 1999 rupture zone (hereinafter referred as near-fault analysis). Particular attention has been devoted to the drainage pattern analysis and to the setting of depositional and tectonic landforms, as well as to the recognition of recent continental deposits, crucial to detect suitable sediments for dating. The 1:25,000-scale geological and geomorphological map of the Düzce fault (Figure 3) shows landforms, related to the erosional and the depositional processes that have been shaping the Almacik range-front, strongly coupled with near-fault tectonic landforms related to fault location and activity.

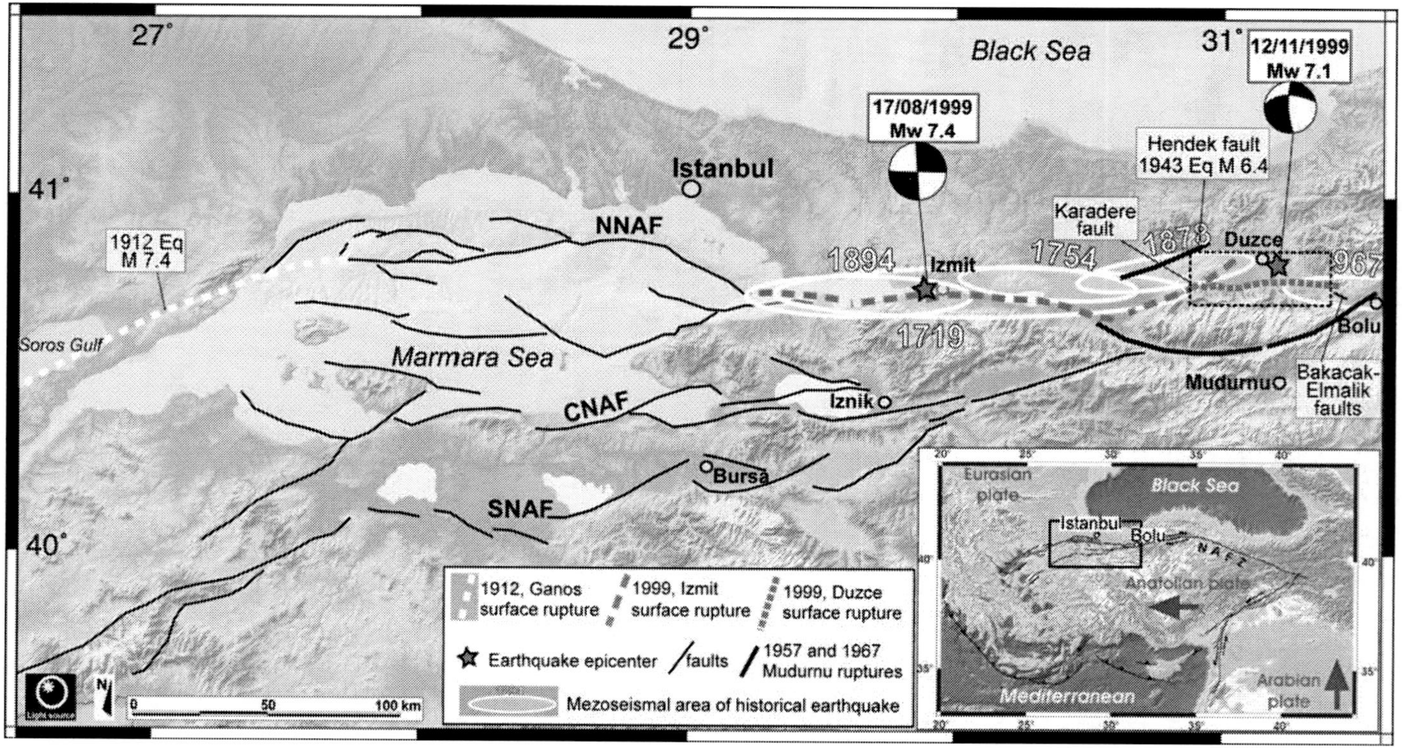

Figure 1. Simplified trace of the North Anatolian fault zone west of the Bolu basin (see lower right inset for location of the area). Portions of the fault zone that ruptured during the 1999 earthquakes are shown. The mezoseismal areas of the historical earthquakes occurred near the Düzce fault segment are shown, too [from Ambraseys and Ambraseys, 2002; Atakan, 2002; King et al., 2001].

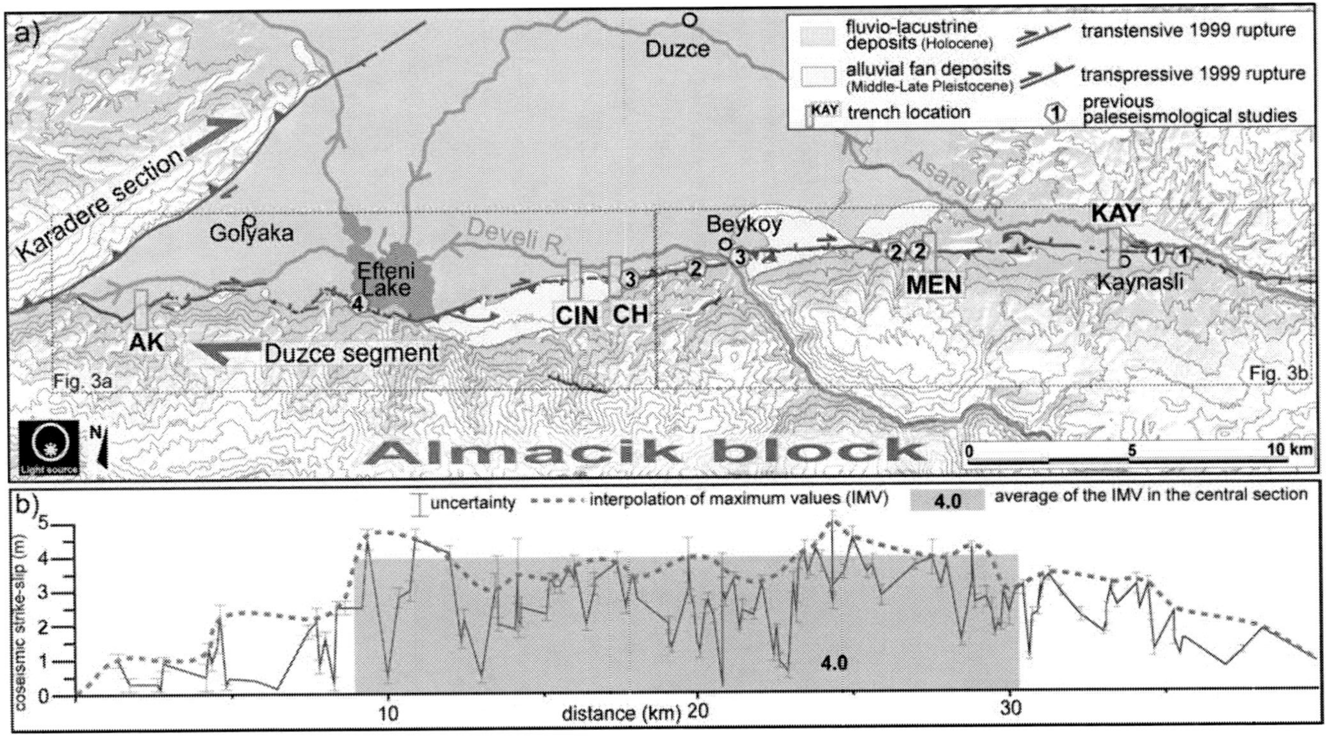

Figure 2. (a) Trace of the 1999 surface ruptures along the Düzce fault (see dashed rectangle in Fig. 1 for location) and location of the five trench sites (rectangles; KAY: Kaynasli; MEN: Mengencik; CH: Cakir Haci Ibrahim; CIN: Cinarli; AK: Aksu). Numbered hexagons are locations where trenching from other authors was performed (1. Hitchcock et al., 2003; 2. Komut, 2005; 3. Emre et al. 2001, 2003a and 2003b; 4. Sugai et al., 2001). Dashed rectangles indicate the location of geomorphic marker analysis. (b) Horizontal slip distribution along the 1999 surface ruptures in the study area.

The erosion of the Almacik block provided important sediment production for the Düzce Basin. These sediments that cover the foot of the range-front are mainly alluvial fans that merge into coalescent aprons (hereinafter referred as bajadas) composed of Middle-Late Pleistocene alluvial fan and Late Pliocene-Early Pleistocene conglomerate-sandstone units (zones x, Figure 3) [Herece and Akay, 2003], and also of active Holocene fans, deposited in subsequent incisions of the old bajadas (e.g., spot y, Figure 3). These generally loose and soft recent deposits recorded persistent landform modification due to faulting and developed well-expressed tectonic-related morphologies whose topography controlled the drainage pattern. Such features are clearly not related to the dynamics of the bajada, since they present tectonic deformations of the original depositional layers. The most common tectonic landforms in the study area are: tilted surfaces, fault escarpments, benches, saddles alignments, linear valleys, shutter ridges, pressure ridges, and sag ponds. This tectonic landforms appear to be the surface expression of a localized slip plane at depth, and the width of the near-fault deformation zones at the surface is related both to the thickness of young alluvial cover sequences and to free surface effects, as observed along other seismogenic structures (e.g., Landers [Sieh et al., 1993; Johnson et al., 1994; Lazarte et al., 1994]; Dast-e Bayaz [Tchalenko and Ambraseys, 1970; Tchalenko and Barberian 1975]). Such tectonic landforms reflect also a small vertical component of deformation, resulting from strike-slip movement along a non rectilinear fault plane that produces transtension at releasing bends (e.g., spots z, Figure 3a) and transpression at restraining bends (e.g., spots q, Figure 3).

Along the western section of the Düzce fault, the 1999 coseismic fault trace bounds the triangular embayments of the range front, that host bajadas. Here, the rupture follows a saw-tooth trajectory (e.g., spots k, Figure 3a) with a few E-W trending elements (e.g., spots w in Figure 3a), associated to subtle and youthful landforms. This suggests that the E-W trending elements adjusted only a small cumulative deformation and may represent an incipient stage of evolution of the fault system from irregular to linear geometry [Pucci et al., 2006 and 2007]. Because the bajada of this section is bounded by the fault to the north, it has a topographic profile with steep fanheads and is characterized by linear and abrupt transitions to the alluvial-lacustrine deposits of the Düzce plain. Linear escarpments, with strong stream incisions across the bajada and suspended terraces can be found at the stream outlets along the range-front foot (spots h, Figure 3a), implying the occurrence of an important subsidence north of the Düzce fault.

Conversely, the eastern section of the Düzce fault (Figure 3b) presents a more regular, localized, prominent E-W trace paralleling the mean trend of the whole

Düzce fault. This E-W trending section crosscuts both recent continental deposits and bedrock and is associated with well developed landforms indicating its persistent long-term activity. It is characterized by shutter and pressure ridges, 300 to 1000 m-long, up to 120 m high (e.g., spots r, Figure 3b), with elongated shapes and long axes paralleling the escarpment elements they are associated with. In many cases, the ridges act as natural dams for northward flowing drainages, being thus responsible for the localization of ponds and the trapping of Holocene fans against the scarp (e.g., spots t, Figure 3b). The tectonic influence on drainage pattern along the fault is also expressed by stream capture. In fact, beheaded channels (i.e., abandoned downstream reaches) flow through fault-parallel linear valleys and connect to the nearby stream, creating apparent sinistral deflections and perched ponding areas (points s and p, respectively, in Figure 4a). Only the easternmost part of this E-W trending section shows subtle near-fault tectonic landforms, going through the Holocene deposits of the Kaynasli pull-apart basin.

Overall, the 1999 displacement zone closely mimics the long-term tectono-morphologic pattern, since the coseismic vertical component of displacement emphasizes the pre-existing relief [see also Pucci et al., 2006]. Geometry, location and style of the coseismic ruptures, strongly indicate that they have been persistent for a long time (several seismic cycles), along the eastern section of the Düzce fault (Figure 3b). Moreover, the geomorphic expression of the Düzce fault testifies that the bulk of the dextral cumulative displacement is localized in a fault zone, that has been experiencing 1999-like surface rupturing for recent, repeated seismic cycles. In particular, the eastern section of the Düzce fault shows the narrowest and best localized deformation of the fault zone [see Pucci et al., 2007] that allows for a more accurate estimate of its long-term offset and that crosscuts recent continental deposits useful for dating [see Pucci et al., 2008].

3.2. The Offset Geomorphic Markers

Along the western part of the eastern fault section, we recognized offset geomorphic markers, such as river terraces and alluvial fans, related to the drainage network evolution at the outlets to the Düzce Basin. These landforms of the study area are suggested to be climatically-controlled and related to the major alluviation episodes (optima) in the Mediterranean region. In fact, rivers

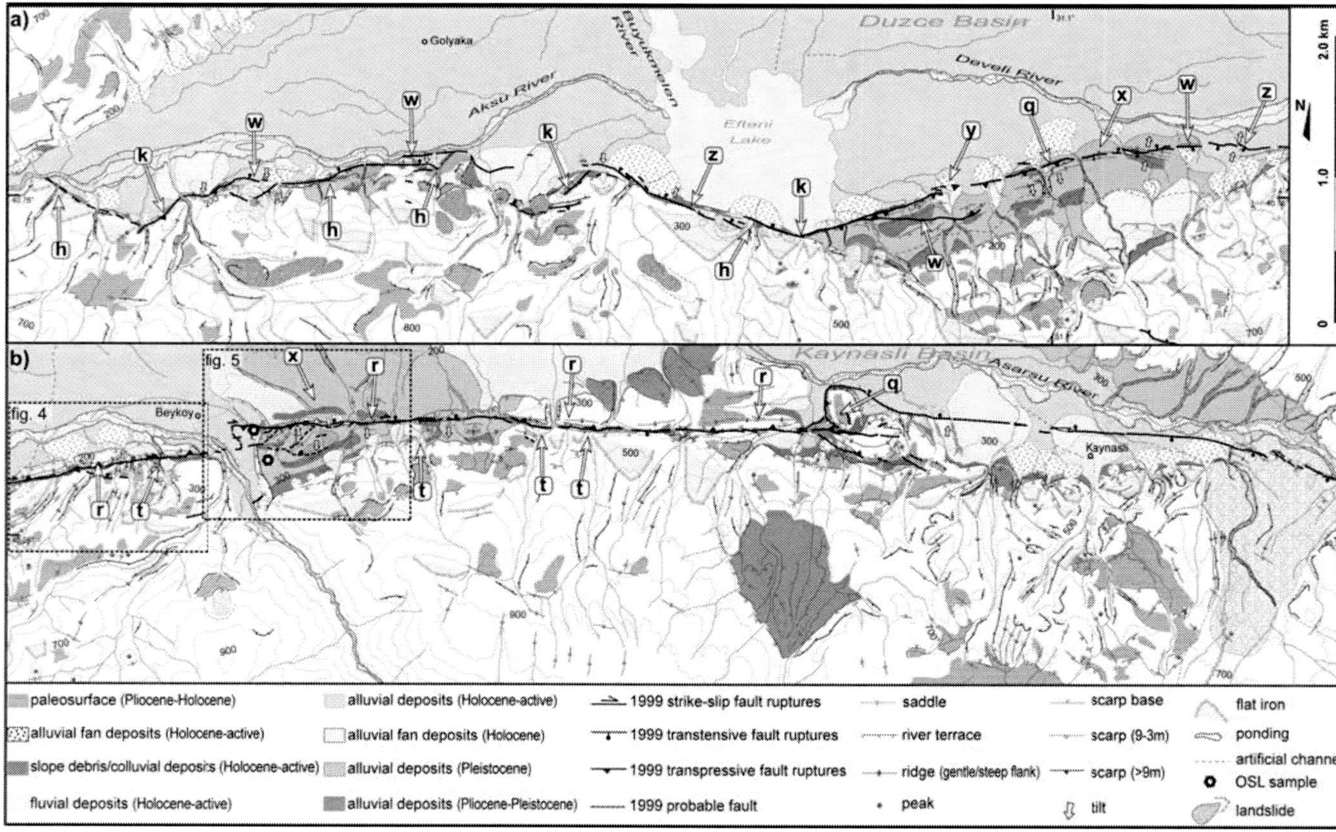

Figure 3. Simplified geomorphological and coseismic fault trace map of the two parts of the Düzce fault. a) Western section. b) Eastern section. Continental deposits from 1:20,000 scale field survey, bedrock from Herece and Akay [2003]. Letters indicate locations discussed in the text. Contour interval 100 m. (see Figure 2a for location).

alluviation phases occurred at the end of dry, cooling trends and the rivers degradational phases at the transition from arid to humid conditions [Pucci et al., 2008]. For these river terraces and alluvial fans we reconstructed the original morphology that has been right-laterally dissected by the Düzce fault motion.

West of Beykoy, we mapped four deflected streams (Figure 4) useful for slip rate calculation assuming a Holocene stream history as follows: (1) the stream was free to flow along a quasi-straight channel and to discharge alluvial deposits with aggradational behavior; (2) because of climatic influence the stream switches to degradational behavior and starts incising within the most recent alluvial surface, the channels remain entrenched and produce terrace remnants [Bull, 1991]; and (3) being unable to change its course, the entrenched stream increases its deflection each time slip occurs on the fault it crosses [Ouchi, 2004, 2005]. At this site, the streams flow north-eastward along entrenched valleys of the range front. When they cross the south-facing fault escarpment, they find as obstacle both bedrock shutter ridges (e.g., spots x, Figure 4a) and remnants of the Holocene alluvial deposits (e.g., spots w, Figure 4a), and turn east to flow along the fault zone (e.g., spots y, Figure 4a). After about a hundred meters, they cut the fault-controlled dam through outlets to the Düzce Basin. The measurement of the deflections of these channels (comparable due to their identical Strahler's [1952] order) yield average cumulative offset of about 106±24 m (Figure 4b). Taking into consideration the assumptions previously explained, in order to establish the timing of the last depositional phase (prior to the stream entrenchment) that likely marks the onset of the stream deflection, we dated the alluvial deposits of the Holocene alluvial remnants. We sampled for radiocarbon dating the top of the remnant of the Holocene alluvial deposits (Figure 4a). The bulk sediment samples provided a calibrated age of of 6720±120 yr BP that represent the onset of the deflection,. This age of the stream deflections translates to a slip rate of 15.7 ± 4 mm/yr for the last ~7000 yrs.

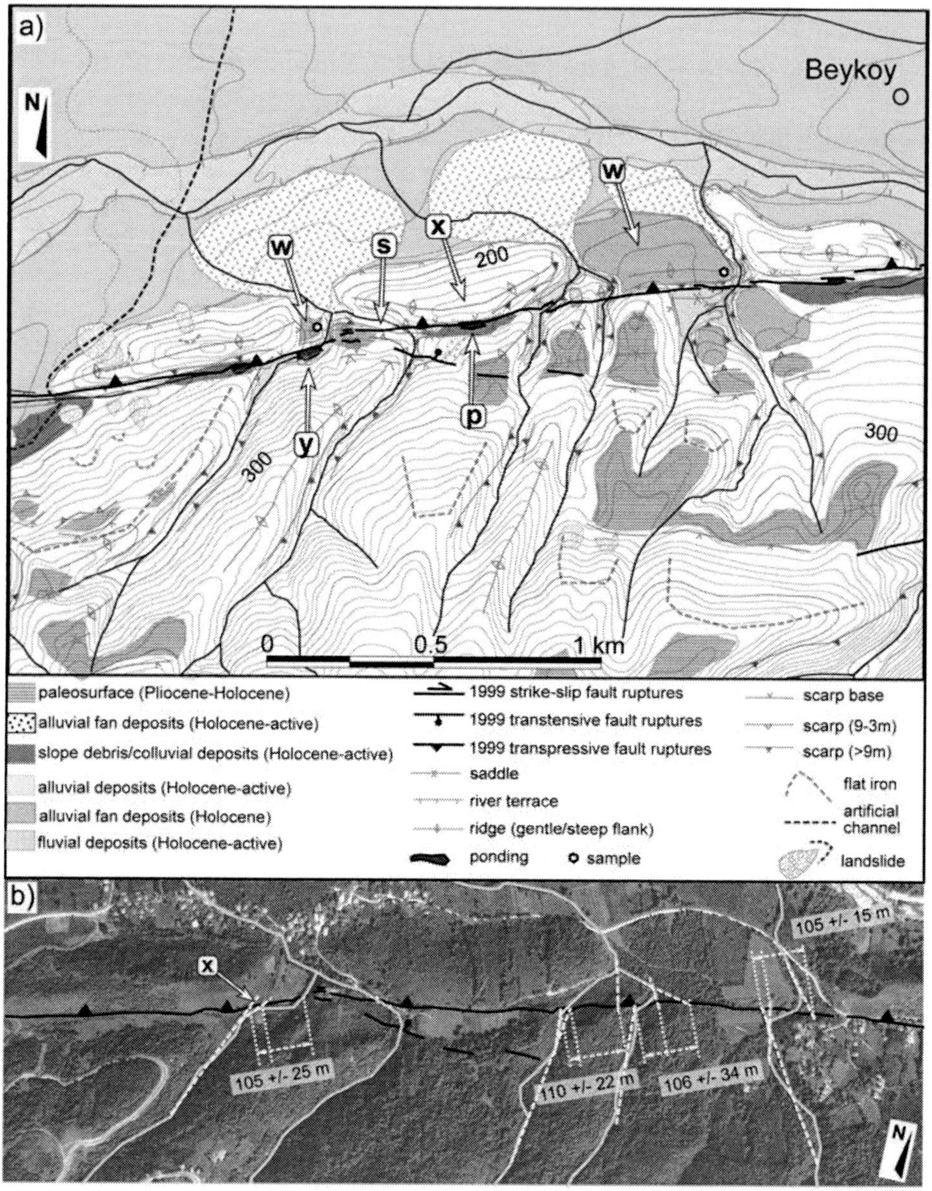

Figure 4. (a) Simplified geomorphological map of a streams offset area. Contour interval 10 m. (see Figure 3b for location). (b) Stream deflection measurements. Dashed lines indicate the possible projections on the fault of the stream channels. Dotted white lines indicate the piercing points used for offset calculations. 1:18,000 scale aerial-photo on the background.

The Develi River is one of the largest water courses of the area (Figure 3). In contrast to the deflected smaller streams, the Develi River is sufficiently large to remove a large volume of sediments as well as any evidence for single coseismic displacement during the interseismic period [e.g., Sugai, 1993]. If this is the case, only features that are not more related to the present dynamics of the streams are suitable evidence for the ongoing deformation (e.g., a flight of abandoned fluvial terraces [Van Dissen and Berryman, 1996]). East of Beykoy, the present-day morphology of the Develi River outlet, shows the presence of remnants of a large Pleistocene alluvial fan, whose sediments have been incised later by the formation of flights of downward-stepping terrace surfaces. At present, the Düzce fault crosscuts and shifts the alluvial fan deposits far away from the path of the river, sheltering from erosion the north-eastern terraces (II, III, IV). The terrace risers (i.e., erosional scarps between the consecutive terraces) observed on both sides of the Düzce fault (II-I, III-II, IV-III) offer suitable piercing points for cumulative offset determinations (Figure 5). The radial fan slope is the oldest and highest depositional surface that decreases downfan with a convex-up profile (IV, Figure 5b). The young terrace surfaces are progressively inset and show a downward-stepping flight arrangement shaped by cut-and-fill alluvial behavior (Figure 3b). The displacement of the III–IV and II–III risers and the resultant morphological juxtaposition were not influenced by any other important stream (i.e., capture) or strong erosional process able to substantially modify them. Following this observation, we correlated risers on both sides of the fault and their tectonic evolution was reconstructed. For the III–IV and II–III order risers, we measured 890±110 m and 300±20 m offsets, respectively (Figure 5a). By means of Optically Stimulated Luminescence method, we dated the top of the alluvial deposits of the terraces, which represent the onset of right-lateral tectonic shift at 60,170 ± 6280 yr BP and 21,700 ± 1850 yr BP. These data translate to slip rates of 14±2.1 mm/yr and 15.2±3.5 mm/yr, for the last ~21 ka and ~60 ka, respectively.

Taking into account that ages derive from samples collected in the upper layers of the alluvial surfaces, while the deformations originate from the measurement of younger erosional features, the slip rate calculated by means of such ages should, therefore, be considered a minimum.

Although the calculated slip rates span ages of 7 to 60 ka BP, these values are similar. Assuming a constant slip rate for the past 60 ka, the mathematical linear regressions that interpolate the maximum and the minimum slip rate intervals yield values of 18.2 and 11.8 mm/yr, respectively. These results translate into a mean slip rate of 15.0 mm/yr±3.2 mm/yr (Figure 6).

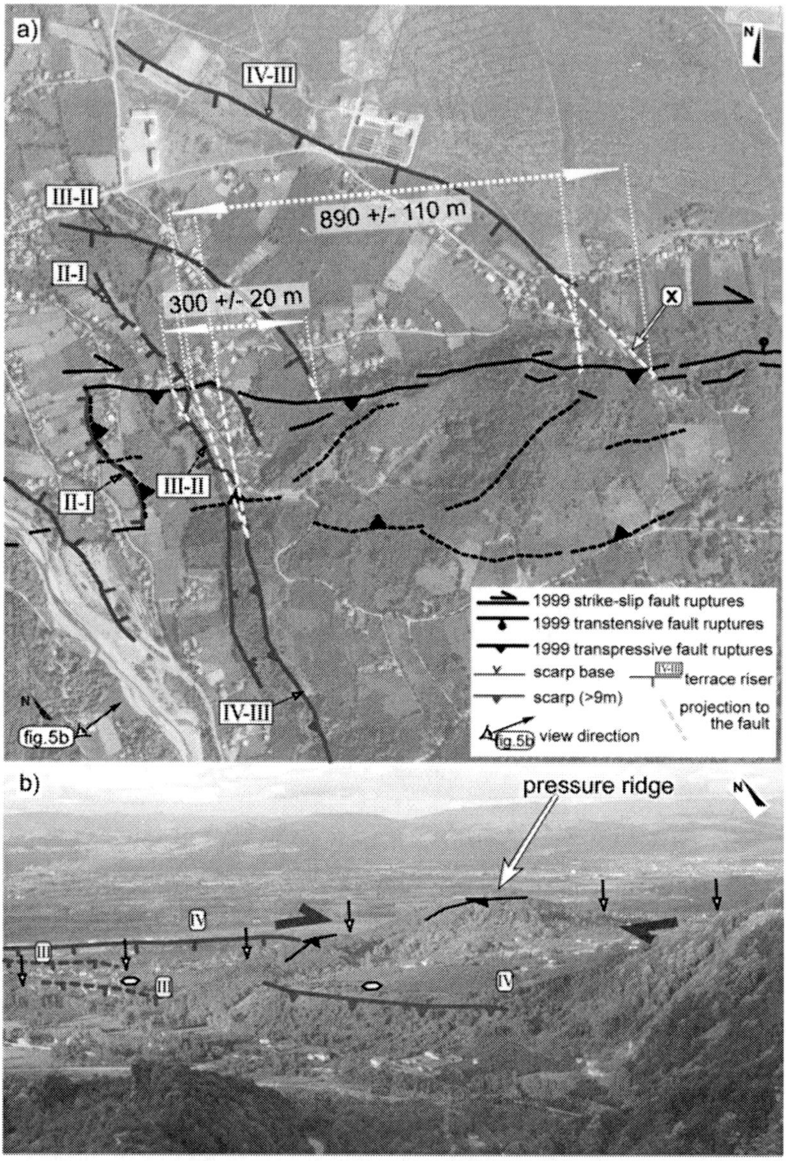

Figure 5. (a) Terrace risers (IV-III and III-II) are severely offset. Dashed lines indicate the possible projections on the fault of the terrace risers. Dotted lines indicate the piercing points used for offset calculations. 1:18,000 scale aerial-photo on the background. (b) Oblique view of the pressure ridge and the terrace risers (see figure a for view direction). Labels III and IV refer to terrace order. Small red arrows point to the coseismic rupture trace. Hexagons locates sampling sites for OSL dating (see Figure 3b for location).

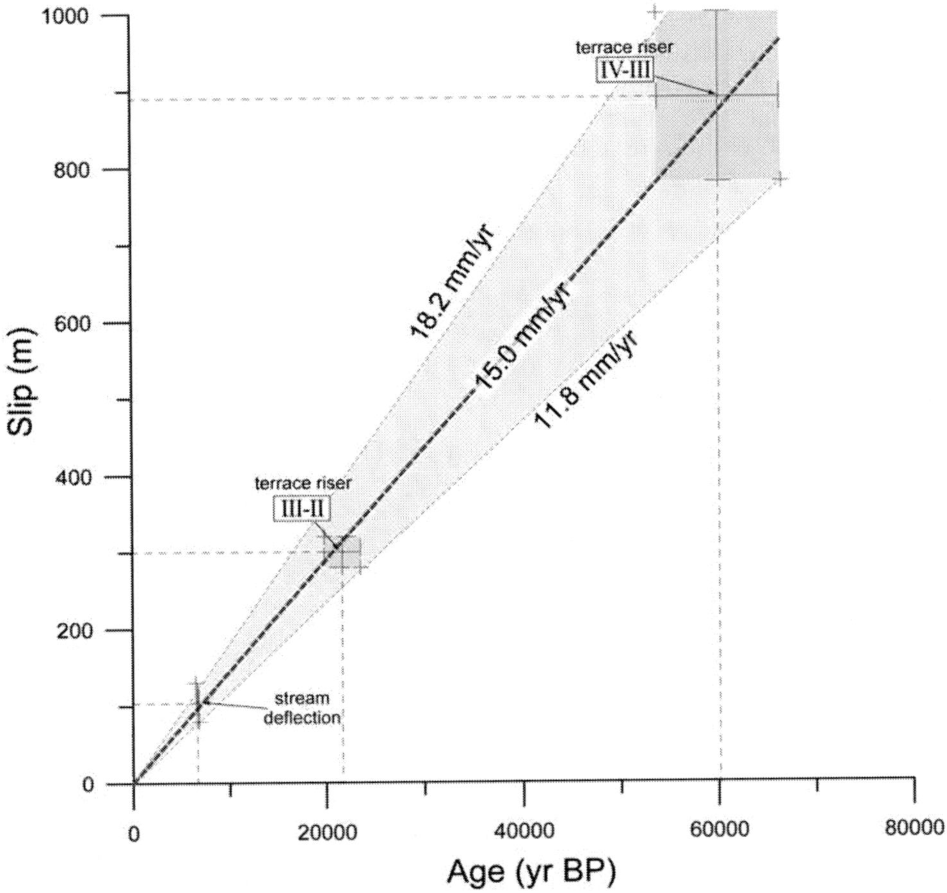

Figure 6. Slip rate diagram of the Düzce fault. The slip rates calculated at different time-scales and for different geomorphic markers (Figure 4 and Figure 5) are reported along with the interpolated uniform slip rate.

The north-flowing drainage system of the Düzce area shows further spectacular right-handed deflections of the streams that cross, at high angle, the Düzce fault. 41 stream deflections, related to cumulative offset of the fault, have been measured (Figure 7a), for the most part of which has not been found any age control to be used for slip-rate estimations. These streams present different Strahler's [1952] order and lithologies of the channel bed, that are supposed to severely influence the stream response to tectonic deflection, mainly by controlling the stream power and the consequent efficiency to remove evidence for coseismic displacements. Figure 7b shows the stream deflection distribution.

On the graph are reported curves (cumulative offset curves) obtained by multiplying the 1999 coseismic right-lateral slip distribution curve (Figure 2b) that represent the cumulative offset that is expected, assuming a "characteristic-earthquake model" [Schwartz, and Coppersmith, 1984], due to the repetition of a different number (N) of 1999-like surface rupturing events. Despite the already-mentioned intrinsic differences, apart from few exceptions, the distribution of the stream deflections appear to mimics the curve given by 20-30 1999-like earthquakes.

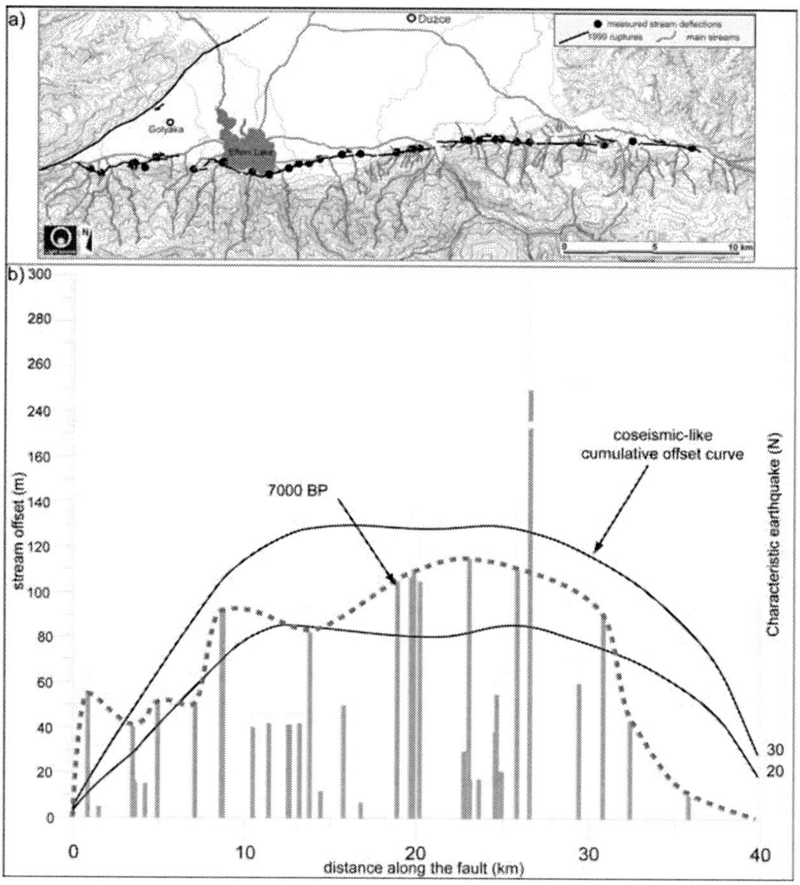

Figure 7. Stream deflection distribution. Bars indicate the stream location and amount of the offset, thin bars indicate uncertainties. Black curves indicate cumulative offset curve obtained by multiplying the 1999 coseismic right-lateral slip distribution (Figure 2b). Thick gray curve is the envelope of the maximum stream offsets.

The period of occurrence of the 20-30 1999-like earthquakes, as like as its relative cumulative offset curve, is indicated by the offset of the streams deflected for ca. 100 m that was radiocarbon dated about 7000 yr BP (Figure 7b). Summarizing, on the basis of the "characteristic-earthquake model" and constant slip rate assumptions, stream deflections across the Düzce fault may be explained by repetition of 20–30 1999-like earthquakes during the last 7000 years, thus the recurrence time for surface rupturing events is 290±60 yr. Notably, the deflection of the streams that crosses the Düzce fault occurred before 7000 BP, at the end of the last deglaciation period (Atlantic, warm and humid climatic conditions [Brocard and Van der Beek, 2006]), indicating the scarce potential of preservation of deflections occurred on previous morphologies that have been re-shaped by last important climatic events.

4. Paleoseismological Data

In order to reconstruct the record of last large earthquakes and provide more constraints for the evaluation of earthquake recurrence on the Düzce fault, 10 trenches were excavated at five sites distributed along the entire surface fault trace (for details see Pantosti et al. [2008]). A major goal was the recognition and dating of individual pre-1999 paleoearthquakes and their possible correlation to known historical earthquakes. Consequently, results from previous paleoseismological investigations were compared along with their implications for the seismic behavior of the Düzce segment of the NAFZ.

On the basis of sedimentary and structural relations, evidence for 2 to 3 surface faulting paleoearthquakes predating the 1999 event in all the trenches were found. Because most of the dates recovered from the trenched deposits indicate young ages, thereby preventing the determination of a precise age, radiocarbon dating was integrated with 210Pb analyses and archaeological information (for details see Pantosti et al. [2008]).

Under the assumption that similarly to the 1999 event, paleoearthquakes on the Düzce fault ruptured the whole fault, on the basis of their age compatibility and their local sequence, events between different trenches were correlated (Figure 8). Merging the radiocarbon results obtained from all the trenches, a seismic history for the Düzce fault for the past 1000–1200 years was proposed. Evidence for the Düzce paleoevents was not found in every trench. As a first approach, in absence of clear stratigraphic constraints, the same sequence of events is assumed to be repeated in each trench. Conversely, when no age compatibility exist a different correlation is proposed.

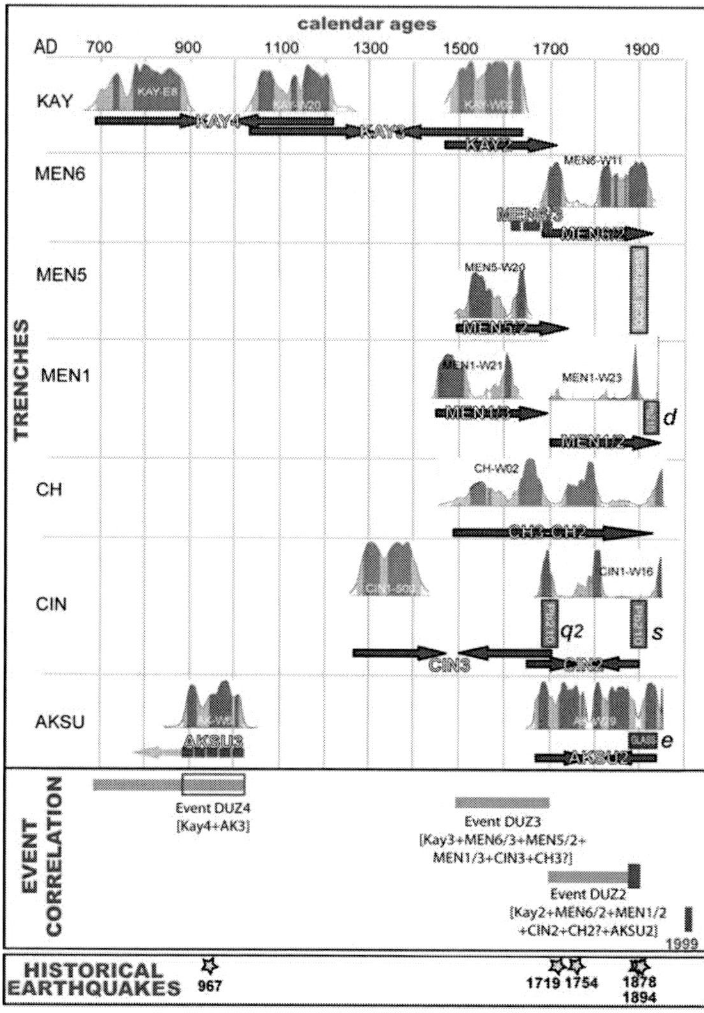

Figure 8. Correlation of paleoearthquakes and inferred age ranges. The upper panel shows the radiocarbon age probability distribution, as well as the 210Pb and the archaeological estimates used to set the age of an event horizon in each trench. Black arrows indicate whether the sample pre-dates or postdates the event. Dashed gray lines indicate a preferred age for the event. The middle panel shows the correlations among trenches. Gray horizontal lines represent the best age range for the event. Dark gray rectangles show preferred ages of events. Correlated events (DUZ2 to DUZ4) represent pre-1999 surface faulting earthquakes that ruptured the same fault extent as in 1999. In the lower panel historical earthquakes (stars) known to have occurred in proximity of the Düzce fault are reported.

The event DUZ2 occurred between A.D. 1700 and 1900. The 210Pb analysis from Cinarli trench (CIN, Figure 8) and the age estimate of the piece of glass from Aksu trench (Figure 8) suggest that DUZ2 occurred close to A.D. 1900. On the basis of radiocarbon dating, the occurrence of DUZ3 is confined during the past 500 years from trenches CIN, KAY and MEN6 (Figure 8), with the youngest age set to circa A.D. 1700 from 210Pb analysis in CIN trench. Limited evidence and age constraints exist for DUZ4 that falls in the interval A.D. 685–1020. On the basis of stratigraphic considerations from Aksu trench the age of occurrence of DUZ4 can be possibly limited to the range A.D. 890–1020.

There are no obvious correlations of these three surface faulting paleoearthquakes that ruptured the Düzce fault before the November 1999 event with the large events reported in the historical catalogues [Ambraseys and Finkel, 1995; Guidoboni et al., 1994; Guidoboni and Comastri, 2005]. The only historical earthquakes that are known to have produced damage near the Düzce fault are the (1) A.D. 967 that was felt in the area of Bolu, (2) A.D. 1719 earthquake that appears to be a perfect twin of the August 1999 Izmit earthquake, (3) A.D. 1754 that is located in the Sapanca area, (4) A.D. 1878 earthquake that is hypothesized on the Hendek fault or in the Akyazi plain, and (5) A.D. 1894, which occurred on the Izmit fault segment farther west [Ambraseys and Finkel, 1995; Ambraseys, 2002, 2006; Atakan et al., 2002; King et al., 2001] (Figure 8; see Figure 1 for earthquake mezoseismal areas). The A.D. 1719, A.D. 1878, A.D. 1894 and A.D. 967 earthquakes are all compatible with the recognized paleoevents of the Düzce fault. However, given the historical damage reports, only the A.D. 1878 and A.D. 967 earthquakes appear to be good candidates for DUZ2 and DUZ4. Given the low density of population and lack of cultural centers in the area, records of DUZ3 (A.D. 1500–1700) may have been lost or overprinted by the disaster produced by the 1719 earthquake farther west.

Previous paleoseismological investigations [Sugai et al., 2001; Emre et al., 2001, 2003a, 2003b; Hitchcock et al., 2003; Komut, 2005] show evidence for surface faulting paleoearthquakes during the past 2000 years or more. By integrating results from all previous works on the Düzce fault, together with those from Pantosti et al. [2008] the record of last large earthquakes was extended up to the past two millennia, specifically through two older events, at ~A.D. 600 and ~A.D. 200, that were described by Sugai et al. [2001] (Figure 9).

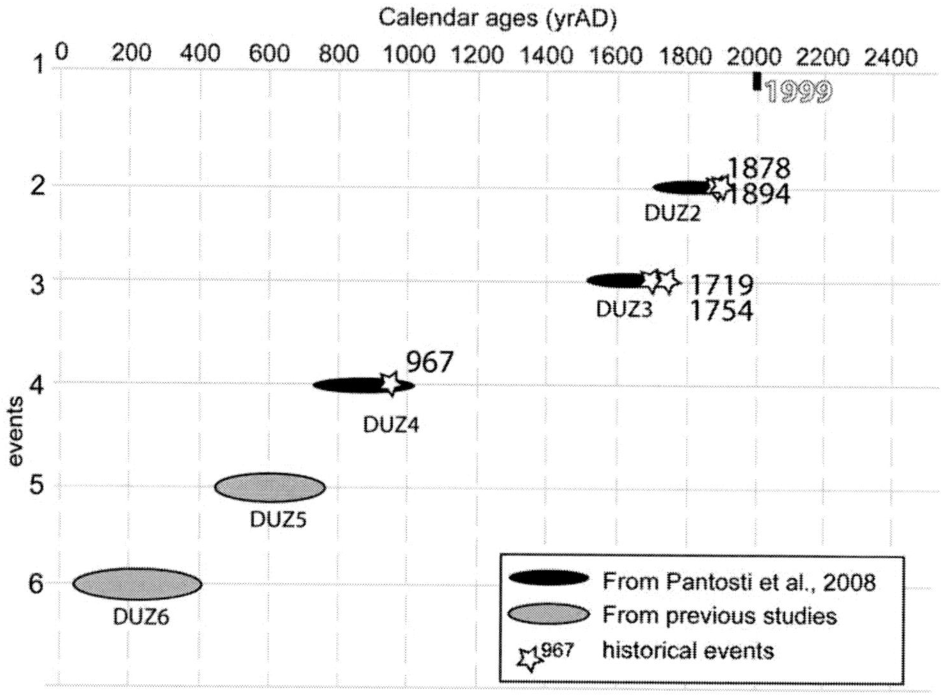

Figure 9. Integration of paleoseismological results from this work and previous ones. Ellipses show the age range uncertainties for paleoearthquakes. Correlated historical earthquakes are shown too (stars). Recent earthquakes appear to have occurred more closely spaced than previous ones.

On the basis of the results from this work, four events since A.D. 685–1020 (possibly A.D. 890–1020) yield an average recurrence for the Düzce fault of 380±50 yr (possibly 350±20 yr). If we include paleoearthquakes from previous studies, which reach farther back in time 2000 years, we obtain a similar figure for the average recurrence of 355±35 yr. However, the three most recent events, including 1999, appear more closely spaced. Average recurrence for the four older events (DUZ3 to DUZ6) can range between 460±100 yr, whereas for the three most recent (1999 to DUZ3) can range between 195±55 yr.

5. COMPARING GEOMORPHOLOGICAL AND PALEOSEISMOLOGICAL RESULTS

With the aim of better comprehending the behavioral model of strain accumulation of the Düzce fault, we performed a comparison between slip rates at short and intermediate time scales, from paleoseismological and geomorphological results, respectively, was performed.

Since it was no possible to measure long-term or single paleoevent offsets from trenches, the paleoseismological investigations can only indirectly offer estimimations on the Düzce fault slip rate over the last two millennia. By assuming the Düzce fault as a segment of the NAFZ defined by persistent segment boundaries and by considering the 1999 coseismic slip as representative for the behavior of this fault (assumption of "characteristic earthquake"), we used the seismic record from paleoseismological study coupled with the 1999 surface coseismic slip, to obtain a first approximative picture of fault slip rate. In order to relate the paleoseismological and the geomorphological data, we assigned to each paleoevent 4.0 m of slip per event, that is the average 1999 coseismic horizontal slip (representative of the fault section from which we studied the geomorphic marker offsets) (Figure 2b). In Figure 10, we graph the past two millennia paleoevents versus the cumulative slip and we constructed the lines that describe the slip rates in order to compare them with those scaled from the geomorphological analysis. Paleoseismological slip rates are presented as averaged (with related uncertainties) both for all and clusters of the paleoevents.

If we take the most conservative average recurrence time, calculated including all the six events that occurred during the past 2 millennia (i.e., 355±35 yr), the 4.0 m average surface coseismic slip yield a slip rate of 10.2–12.5 mm/a. This short-term slip rate derived from the 1999 maximum slip overlaps with the lower part of that of 11.8–18.2 mm/yr obtained from long-term offset by means of geomorphic markers analysis (Figure 10). The 10 mm/a slip rate derived from GPS measurements [Ayhan et al., 1999, 2001] falls just below the lower value inferred from paleoseismological data. All these data imply that the Düzce fault substantially accommodates, along with the Mudurnu section to the south, the 20 mm/yr of motion along the NAFZ in this area.

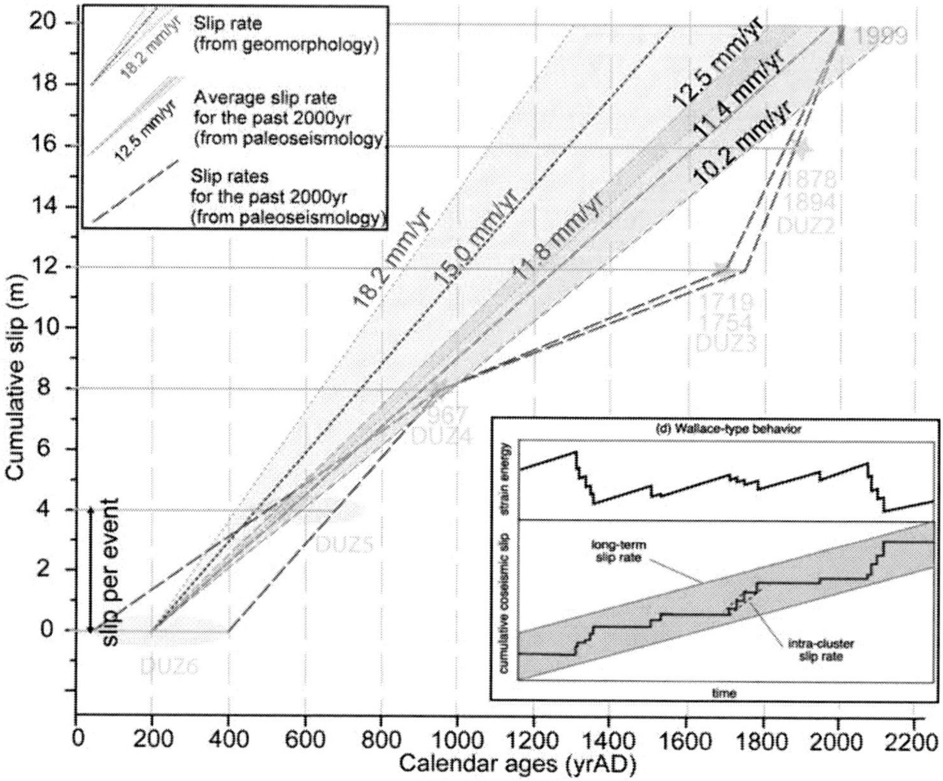

Figure 10. Comparison between slip rates from paleoseismological and geomorphological results. Shaded areas show average slip rate uncertainties for both dataset. Light gray ellipses and stars show the seismic record of the last two millennia. 1999-like slip per event is assigned. Dashed lines indicate the short-term slip rate fluctuations suggested by the paleoevents. The inset shows the Wallace-type strain release model [1987] (modified from Friedrich et al. [2003]).

Should we consider the paleoseismic record to be complete, the sequence depicted in Figure 9 would be suggestive of earthquake clustering in time. A bimodal paleoearthquakes distribution indicates fluctuations during the past two millennia, with clustered events separated by a longer interseismic interval (967-1719/1754 AD). Since the interseismic interval is not perfectly periodic, the cluster distribution of the paleoearthquakes produces a bimodal slip rate that suggests strain release of the Düzce fault, presenting fluctuations, with clustered high strain release and uniform, low strain release (Figure 10). Over long time intervals, the strain accumulation should be constant and, in the long-term, strain accumulation (from geomorphic markers analysis) and strain release (from paleoseismology) should be the same. In this case, the clustered strain release would recover the small amount of strain energy released before 1700 A.D., in order to keep constant the short-term slip rate, as exemplified by Wallace-type models (see inset in Figure 10).

5. Discussion and Conclusions

Long-term slip rate estimates of the fault, provided through geomorphological mapping and analysis of fault-related cumulative landforms, were used to calculate recurrence time of the Düzce fault and for a comparison with independent results made available by paleoseismological trenching.

It is critical to determine the timing and distribution of individual faulting events because long-term average slip rates may give grossly incorrect assessments of the hazard potential of a single fault. In addition, evaluation of the characteristics of the entire fault system that contribute to the plate margin deformation, providing a combination of regional and detailed information, is essential for an adequate hazard assessment.

Along the western part of the eastern Düzce fault section, the narrowest and best localized deformation of the fault zone allowed for a more accurate estimate of long-term displacements. Here, offset geomorphic markers, such as river terraces and alluvial fans, were recognized and were used to calculate mean slip rate of 15.0 mm/yr±3.2 mm/yr for the last 60 ka BP. Furthermore, the distribution of spectacular right-handed stream deflections induced by Holocene cumulative offset of the fault has been estimated to be produced by 20–30 1999-like earthquakes. On the basis of the "characteristic-earthquake model" and constant slip rate assumption, stream deflections across the Düzce fault may indicate a recurrence time of 290±60 yr for surface rupturing events.

Under the same assumption of a "characteristic-earthquake model," the correlation of paleoearthquakes between different trenches along the Düzce fault and the integration of results from previous works show a record of six large earthquakes for the past two millennia. From these paleoearthquakes, we obtain an average recurrence time for large earthquakes of 355±35 yr.

The overall average recurrence time for the past two millennia calculated from displaced geomorphic markers and paleoseismological analyses are coherent, presenting ranges of values whose overlap is confined between 320 and 350 years.

Slip rate derived from the average recurrence time of the paleoevents and from the average 1999 coseismic slip of 4.0 m, as slip per event, allows us to estimate, in first approximation, an average slip rate of 11.4 ± 1.1 mm/yr. Comparison between average slip rates from paleoseismological and geomorphological results are comparable, since they present ranges of values overlapping around 12.0 mm/yr.

A detailed analysis of the paleoearthquakes highlights a cluster distribution that produces a bimodal slip rate. This is suggestive of strain release presenting fluctuations in time, with low strain release periods recovered by clustered high strain release periods, indicating a Wallace-type strain release model for the Düzce fault.

Increasing the geological observations is crucial to answer one of the most important questions regarding the significance of the measurements of active surface deformation in the continental lithosphere: what is the relationship between strain accumulation and strain release? On short time scales (years), both strain accumulation and release are monitored using high-precision space-geodetic techniques (GPS; InSAR, etc.). However, on longer time scales (e.g., thousands of years, duration of interseismic intervals), strain accumulation measurements are derived only from geologic strain release data as proxies, and more detailed investigations are needed to characterize slip rates and the timing and patterns of faulting events. Near-fault geodetic, paleoseismological, and geological estimates of fault slip rates have been observed to differ (e.g., Basin and Range Province [Friedrich et al. 2003]). In fact, as shown by the Düzce fault, the average slip rates may remain constant, but variations exist on small time scales larger than the earthquake cycle [Wallace, 1987]. Plate margins, such as the NAFZ along the study area, could present irregularities in crustal thickness, changes in lithology, local variations of the regional stress systems and complex structural style of the active fault system originated from the tectonic inheritance. In these regions, non-uniformity of occurrence of large slip events on seismogenic faults and small time scale variations of slip rates are probably a normal characteristic of faulting.

REFERENCES

Akyüz, H.S., A.A. Barka, E. Altunel, R.D. Hartleb, and G. Sunal (2000), Field observations and slip distribution of the November 12, 1999 Düzce earthquake (M=7.1), Bolu - Turkey. In: A.A. Barka et al. (Eds.), *The 1999 Izmit and Düzce earthquakes; preliminary results*. Istanbul Technical University, Istanbul, Turkey, pp. 63-70.

Akyüz, H.S., R.D. Hartleb, A.A. Barka, E. Altunel, G. Sunal, B. Meyer and R. Armijo (2002), Surface rupture and slip distribution of the 12 November 1999

Düzce earthquake (M7.1), North Anatolian Fault, Bolu, Turkey, *Seism. Soc. Am. Bull.*, 92 (1), 61-66.

Ambraseys, N.N, (2002), The seismic activity of the Marmara Sea region over the last 2000 years, *Seism. Soc. Am. Bull.*, 92 (1), 1-18.

Ambraseys, N. N. (2006), Comparison of frequency of occurrence of earthquakes with slip rates from long-term seismicity data: The cases of Gulf of Corinth, Sea of Marmara and Dead Sea Fault Zone, *Geophys. J. Int.*, 165, 516– 526.

Ambraseys N.N. and C.F. Finkel (1995). *The seismicity of Turkey and adjacent areas: a historical review, 1500–1800*, Muhittin Salih Eren, 240 pp., Istanbul.

Armijo, R., B. Meyer, A. Hubert and A.A. Barka (1999), Westward propagation of the North Anatolian fault into the northern Aegean: Timing and kinematics, *Geology*, 27 (3), 267-270.

Ayhan, M.A., C. Demir, A. Kilicoglu, I. Sanli, and S.M. Nakiboglu (1999), *Crustal motion around the western segment of the north Anatolian fault zone: geodetic measurements and geophysical interpretation*. International Union of Geodesy and Geo-physics (IUGG99), Birmingham, United Kingdom, 18-30 July.

Ayhan, M.E., R.B¨urgmann, S. McClusky, O. Lenk, B. Aktug, E. Herece and R.E. Reilinger, 2001. Kinematics of the Mw = 7.2, 12 November 1999, Düzce, Turkey earthquake, *Geophys. Res. Lett.*, 28 (2), 367-370.

Atakan, K., A. Ojeda, M. Meghraoui, A.A. Barka, M. Erdik and A. Bodare (2002), Seismic hazard in Istanbul following the 17 August 1999 Izmit and 12 November 1999 Duzce earthquakes, *Seism. Soc. Am. Bull.*, 92 (1), 466-482.

Barka, A.A., 1992. The North Anatolian fault zone. *Ann. Tecton.* 6, 164–195.

Brocard, G.Y, and P.A. Van der Beek (2006), Influence of incision rate, rock strength, and bedload supply on bedrock river gradients and valley-flat widths; field-based evidence and calibrations from western Alpine rivers (southeast France), Special Paper—*Geological Society of America*, 398, 101-126.

Bull,W.B., 1991. Geomorphic Responses to Climatic Change. Oxford University Press, London, UK, p. 326.

Canıtez, N. and B. Üçer (1967), Computer determinations for the fault plane solutions in and near Anatolia, *Tectonophysics*, 4, 235-244.

Christie-Blick, N. and K.T. Biddle (1985), Deformation and basin formation along strike-slip faults, in: Strike-slip Deformation, Basin Formation and Sedimentation, K.T. Biddle and N. Christie-Blick (Editors), *Soc. Econ. Paleont. Min. Spec. Publ.*, 1-34.

Dewey, J.F., M.R. Hempton, W.S.F. Kidd, F. Saroglu and A.M.C. Şengör (1986), Shortening of continental lithosphere; the neotectonics of eastern Anatolia, a young collision zone, *Geological Society Spec. Pub.*, 19, 3-36.

Emre, Ö., T.Y. Duman, S. Toda, M. Okuno, A. Dogan, S. Ozalp, H. Tsutsumi, F. Tokay, T. Haraguchi, H. Kondo, N. Subito and T. Nakamura (2001), Paleoseismologic Findings on the Düzce Fault: North Anatolian Fault Zone, NW Turkey, American Geophysical Union, Fall Meeting 2001, abstract.

Emre, Ö., Y. Awata and T. Duman (Eds.) (2003a), *Surface rupture associated with the 17 August 1999 Izmit earthquake*. General Directorate of Mineral Research and Exploration, Ankara, Turkey, ISBN: 975-6595-53-1, pp. 29–271.

Emre, Ö., S. Toda, T.Y. Duman, T. Sugai, A. Dogan, Y. Awata, M.Okuno, H. Tsutsumi, S. Ozalp and F. Tokay (2003b), *Recurrence of the large earthquakes on the 1999 Izmit and Düzce surface ruptures, North Anatolian Fault, Turkey*. EGS-AGU-EUG Joint Assembly, Abstracts, 6-11 April 2003, Nice, France.

Friedrich, A.M., B.P. Wernicke, N.A. Niemi, R.A. Bennett and J.L. Davis (2003), Comparison of geodetic and geologic data from the Wasatch region, Utah, and implications for the spectral character of Earth deformation at periods of 10 to 10 million years, *J. of Geophys. Res.*, vol. 108, B4, 2199, doi:10.1029/2001jb000682,

Guidoboni, E., and A. Comastri (2005), Catalogue of Earthquakes and Tsunamis in the Mediterranean Area from the 11th to the 15th Century, 1037 pp., Storia Geofis. Ambiente, Bologna, Italy.

Guidoboni, E., A. Comastri, and G. Traina (1994), Catalogue of Ancient Earthquakes in the Mediterranean Area Up to the 10th Century, 504 pp., Storia Geofisica Ambiente, Bologna, Italy.

Herece, E., and E. Akay (2003), 1:100.000 geological maps of the North Anatolian Fault, General Directorate of Mineral Research and Exploration, Ankara, Appendix 3 and 4.

Hitchcock, C., E. Altunel, A.A Barka, J. Bachhber, W. Lettis, Ö. Kozci, J. Helms and S. Lindvall (2003), Timing of Late Holocene earthquakes on the eastern Düzce fault and implications for slip transfer between the southern and the northern strands of the North Anatolian Fault System, Bolu, Turkey, *Turkish J. Earth Sci.*, 12 (1), 119-136.

Jackson, J. and D. McKenzie (1988). The relationship between plate motions and seismic moment tensors, and the rates of active deformation in the Mediterranean and Middle East, *Geoph. J. Int.*, 93 (1), 45-73.

Johnson, A. M., R. W. Fleming, and K. M. Cruikshank (1994), Shear zones formed along long, straight traces of fault zones during the 28 June 1992 Landers, California, earthquake, *Bull. Seism. Soc. Am*, 84, 499–510.

Kahle H. G., M. Cocard, Y. Peter, A. Geiger, R. Reilinger, S.C. McClusky, R. King, A. Barka and G. Veis (1999), GPS-derived strain rate field within the boundary zones of the Eurasian, African, and Arabian Plates, *J. Geophy. Res.* 105 (B3), 23 353-23 370.

Kahle H. G., M. Cocard, Y. Peter, A. Geiger, R. Reilinger, A. Barka and G. Veis (2000), GPS-derived strain rate field within the boundary zones of the Eurasian, African, and Arabian Plates, *J. Geophy. Res.*, 105 (B3), 23 353-23 370.

King, G.C.P., A. Hubert-Ferrari, S.S. Nalbant, B. Meyer, R. Armijo and D.D. Bowman (2001), Coulomb interactions and the 17 August 1999 Izmit, Turkey earthquake, Comptes Rendus de l'Academie des Sciences, Serie II. Sciences de la Terre et des Planetes, 333 (9), 557-569.

Komut T., (2005), Paleoseismological studies on Düzce fault and geological data on the seismogenic sources in the vicinity of Düzce area, Kandilli Observatory and Earthquake Research Institute Boğaziçi University, Ph.D. Thesis, pp. 155.

Lazarte, C.A., J.D. Bray, A.M. Johnson, and R.E. Lemmer (1994), Surface breakage of the 1992 Landers earthquake and its effects on structures, *Bull. Seism. Soc. Am*, 84, 547–561.

McClusky, S.C., A. Balassanian, A.A. Barka, C. Demir and S. Ergintav (2000). Global positioning system constrain on plate kinematics and dynamics in the eastern Medditerranean and Caucasus, *J. Geophy. Res.*, 105 (B3), 5695-5720.

McKenzie, D.P., (1972), Active tectonics of the Mediterranean region, *Geophys. J. R. Astron. Soc.*, 30, 109-185.

Okay, A.I., E. Demirbag, H.N. Kurt, javascript: do_literal('AU=(Okay Nilgun)');and I. Kuşçu (1999), An active, deep marine strike-slip basin along the North Anatolian Fault in Turkey, *Tectonics*, 18 (1), 129-147.

Ouchi, S., 2004. Flume experiments on the horizontal stream offset by a strike-slip fault. *Earth Surf. Processes Landf.* 29, 161–173.

Ouchi, S., 2005. Development of offset channels across the San Andreas fault. *Geomorphology*. 70, 112–128.

Pantosti, D., S. Pucci, P.M. De Martini, N. Palyvos, G. D'Addezio and C. Zabci (2008), Paleoseismological investigation along the Duzce segment of the North Anatolian Fault Zone (Ms 7.1, Nov. 1999), *Journal of Geophysical Research-Solid Earth*, 113, B01309, doi:10.1029/2006JB004679.

Parsons, T., S. Toda, R.S. Stein, A. Barka and J.H. Dieterich (2000), Heightened odds of large earthquakes near Istanbul: an interaction-based probability calculation, *Science*, 288, 661–664.

Pucci, S., N. Palyvos, C. Zabci, D. Pantosti (2006): Coseismic ruptures and associated tectonic landforms along the Duzce segment of the North Anatolian Fault Zone (Ms 7.1, Nov. 1999), *Journal of Geophysical Research-Solid Earth*, 111, B06312, doi:10.1029/2004JB003578.

Pucci, S., D. Pantosti, M. Barchi and N. Palyvos (2007), A Complex seismogenic shear zone: the Düzce segment of the North Anatolian Fault Zone (Turkey), *Earth Planetary Science Letters*, 262, 185–203, doi:10.1016/j.epsl. 2007. 07.038.

Pucci, S., P.M. De Martini and D. Pantosti (2008): Preliminary slip rates of the Duzce segment of the North Anatolian Fault Zone from offsets geomorphic markers, *Geomorphology*, 97, 538–554.

Reilinger, R.E., S.C. McClusky, M.B. Oral, W. King and M.N. Toksöz (1997), Global Positioning, System measurements of present-day crustal movements in the Arabian-Africa-Eurasia plate collision zone. *J. Geophy. Res.* 102 (B5), 9983-9999.

Reinecker, J., O. Heidbach, M. Tingay, P. Connolly, and B. Müller (2004), The 2004 release of the World Stress Map (available online at www.world-stress-map.org).

Şaroğlu, F., Ö. Emre and İ. Kuşçu (1992). Active fault map of Turkey. General Directorate of Mineral Research and Exploration, Ankara.

Schwartz, D.P., and K.J. Coppersmith (1984), Fault behavior and characteristic earthquakes; examples from the Wasatch and San Andreas fault zones, *J. Geophys. Res.*, 89 (7B), 5681-5698.

Şengör, A.M.C., (1979), The North Anatolian transform fault: Its age, offset and tectonic significance, *J. Geol. Soc. London*, 136, 269-282.

Şengör, A.M.C., N. Görür and F. Şaroğlu (1985), Strike-slip faulting and related basin formation in zones of tectonic escape; Turkey as a case study, In Biddle, K.T., and N. Christie-Blick (Eds.) Strike-slip Deformation, Basin Formation and Sedimentation, *Soc. Econ. Paleont. Min. Spec. Publ.*, 37, 227-264.

Şengör, A.M.C., O. Tüysüz, C. İmren, M. Sakınç, H. Eyidoğan, N. Görür, X. Le Pichon and C. Rangin (2005), The North Anatolian Fault: a new look, *Annu. Rev. Earth Planet. Sci.*, 33, 37-112.

Sieh, K.L., L. Jones, E. Hauksson, K. Hudnut, D. Eberhart-Phillips, T.H. Heaton, S. Hough, K. Hutton, H. Kanamori, A. Lilje, S. Lindvall, S.F. McGill, J. Mori, C. Rubin, J.A. Spotila, J. Stock, H.K. Thio, J. Treiman, B. Wernicke, and J. Zachariasen (1993), Near-field investigations of the Landers earthquake sequence, April to July, 1992, *Science*, 260, 171–176.

Strahler, A.N., (1952), Dynamic basis of geomorphology, *Geol. Soc. Am. Bull.*, 63, 923-938.

Straub, C., H. G. Kahle, and C. Schindler (1997), GPS and geologic estimates of the tectonic activity in the Marmara Sea region, NW Anatolia, *J. Geophys. Res.*, 102 (B12), 27 587-27 601

Sugai, T., 1993. River terrace development by concurrent fluvial processes and climate change. *Geomorphology*. 6, 243–252.

Sugai, T., Y. Awata, S. Toda, Ö. Emre, A. Dogan, S. Ozalp, T. Haraguchi, H. Kinoshita, K. Takada, and M. Yamaguchi (2001), Paleoseismic investigation of the 1999 Du¨zce earthquake fault at Lake Efteni, North Anatolian fault system, Turkey, Annu. Rep. Active Fault Paleoearthquake Res. 1, Active Fault Res. Cent., Tsukuba, Japan.

Tchalenko, J.S. and N.N. Ambraseys (1970), Structural analysis of the Dasht-e Bayaz (Iran) earthquake fractures, *Geol. Soc. Am. Bull.*, 81, 1, 41-59.

Tchalenko, J.S., and M. Berberian (1975), Dasht-e Bayaz fault, Iran: earthquake and earler related structures in bedrock, *Geol. Soc. Am. Bull.*, 86, 703–709.

Van Dissen, R., and K.R. Berryman (1996), Surface rupture earthquakes over the last ~1000 years in the Wellington region, New Zealand, and implications for ground shaking hazard, *J. Geophys. Res.*, 101, B3, 5999-6019.

Wallace, R. E. (1987), Grouping and migration of surface faulting and variation in slip rates on faults in the Great Basin province, *Bull. Seismol. Soc. Am.*, 77, 868–877.

Westaway, R., (1994), Present-day kinematics of the Middle-East and Eastern Mediterranean, *J. Geophys. Res*, 99, 12 071-12 090.

Wong, H K, T. Luedmann, A. Ulug and N. Gorur (1995), The Sea of Marmara; a plate boundary sea in an escape tectonic regime, *Tectonophysics*, 244 (4), 231-250.

In: Recent Progress on Earthquake Geology
Editor: Pierpaolo Guarnieri, pp. 109-145
ISBN: 978-1-60876-147-0

Chapter 5

IN SEARCH OF TSUNAMI DEPOSITS ALONG THE EASTERN COAST OF SICILY (ITALY): THE STATE OF THE ART

M.S. Barbano[1], P.M. De Martini[2], D. Pantosti[2], A. Smedile[2], P. Del Carlo[3], F. Gerardi[1], P. Guarnieri[1] and C. Pirrotta[1]

[1] Dipartimento di Scienze Geologiche, Università di Catania, Corso Italia 55, 95129 Catania, Italy
[2] Istituto Nazionale di Geofisica e Vulcanologia, Sezione Sismologia e Tettonofisica, Via di Vigna Murata 605, 00143 Roma, Italy
[3] Istituto Nazionale di Geofisica e Vulcanologia, Sezione di Pisa, via della Faggiola 32, 56126 Pisa, Italy

ABSTRACT

Eastern Sicily has been affected in historical times by large earthquakes followed by devastating tsunamis, such as the 1169, 1693 and 1908 events.

In order to provide a long-term assessment for tsunami recurrence and related hazards, we developed a multi-disciplinary study, with a paleoseismological approach, aimed to recognize and date historical and paleotsunami deposits.

Starting from information on the effects of known tsunamis (hit localities, inundation areas, run-up heights) and with a geomorphological approach, we selected several sites—such as coastal lakes, marshes and lagoons—potentially suitable for preserving tsunami deposits. In these sites, 64 test gouge cores have been dug by hand and engine coring.

In order to reconstruct paleoenvironments and to identify potential paleotsunami deposits, sedimentological and paleontological analyses were carried out. Magnetic and X-ray analyses were used to highlight susceptibility variations and peculiar small-scale sedimentary structures not detectable through the standard stratigraphic investigation. Moreover, radiocarbon dating and tephra identification provide age ranges of the tsunami deposits and constraints for sedimentation rates allowing the correlation with historical events.

At Capo Peloro in northeastern Sicily, combining archaeological, historical, and C14 data, we associated two tsunami deposits with the earthquakes occurred in 1783 and 17 A.D. We also collected evidence for the occurrence of multiple inundations at sites in the eastern flank of Mt. Etna: three events in the past 580 yrs at the Anguillara site and four events in the past 4000 yrs at the Gurna site. In southeastern Sicily, in the Augusta bay, combining historical, tephrostratigraphical and C14 dating, we reconstructed a tsunami inundation history composed of six events in the past 4000 yrs; the two most recent ones are related to the 1693 and 1169 earthquakes.

Keywords: Tsunami deposits, sedimentation processes, micropaleontology, tephra, environmental analyses, off-fault paleoseismology, coastal hazards

1. Introduction

Geological records are increasingly being used to contribute to knowledge of past earthquakes and to provide estimates of the their probable occurrence, location and magnitude in the future (e.g., Pantosti et al., 2003 and papers therein). As a result of the increased worldwide awareness of coastal hazards, tsunami geology has recently received attention also in regions such as the Mediterranean because of the density of population, infrastructures and economic activities that concentrate along the coasts (e.g., De Martini et al., 2003; Ruiz et al., 2005; Pantosti et al., 2008; Scheffers et al., 2008)

The active tectonic processes in Italy express the ongoing deformation through several important seismic sources responsible for earthquakes that occasionally locally generated tsunamis. However, Italian coastal areas are also exposed to tsunami waves originating from distant sources, such as those belonging to the Aegean subduction zone (e.g., the 365 A.D. Crete earthquake and the ca. 1500 B.C. Santorini plinian eruption).

Reliable Italian historical records of important inundation of the coastland cover about 2000 yrs. Eastern Sicily has been affected in historical times by large earthquakes (Figure 1) that, as described in the historical reports (Table 1), were followed by devastating tsunamis such as the 1908, 1783, 1693, and 1169 earthquakes.

Although a significant number of large-magnitude tsunamis are reported for the Italian region (Caputo and Faita, 1984; Tinti et al., 2001, 2004), a few investigations of the geological records of tsunamis have been undertaken along the Italian coasts. De Martini et al. (2003) identified tsunami deposits in the Gargano area; in the same region, Gianfreda et al. (2001) described the presence of three wide washover fans in the Lesina coastal barrier formed by three distinct tsunamis that occurred in historical times. Mastronuzzi and Sansò (2000; 2004) and Mastronuzzi et al. (2007) explained boulders of Pleistocene calcarenites, up to 80 t in weight scattered along the Ionian coast of Apulia, as related to three tsunamis that occurred between the fifteenth and eighteenth centuries. Scicchitano et al. (2007) interpreted calcareous boulders between the towns of Augusta and Siracusa (Sicily) scattered along wide terraces, 2–5 m a.s.l., as detached and transported by historical tsunamis.

During the past years, we developed a multi-disciplinary study aimed at the recognition and dating of historical and paleotsunami deposits in eastern Sicily in order to provide new information on the extent and frequency of tsunami inundation in the investigated areas.

Starting from a detailed analysis of coeval accounts, we have compiled a database storing information on the effects of known tsunamis (hit localities, maximum run-up, inundation areas, etc.). To select areas that can have been likely invaded by tsunami waves in the past, these data were used together with a geomorphologic approach, which took into account also past centuries' coastline changes. Coring was performed at several sites and the collected data were studied in detail to distinguish deposits of high-energy environments transported by tsunami waves in the local sediments of the site, generally deposited in a low-energy environment.

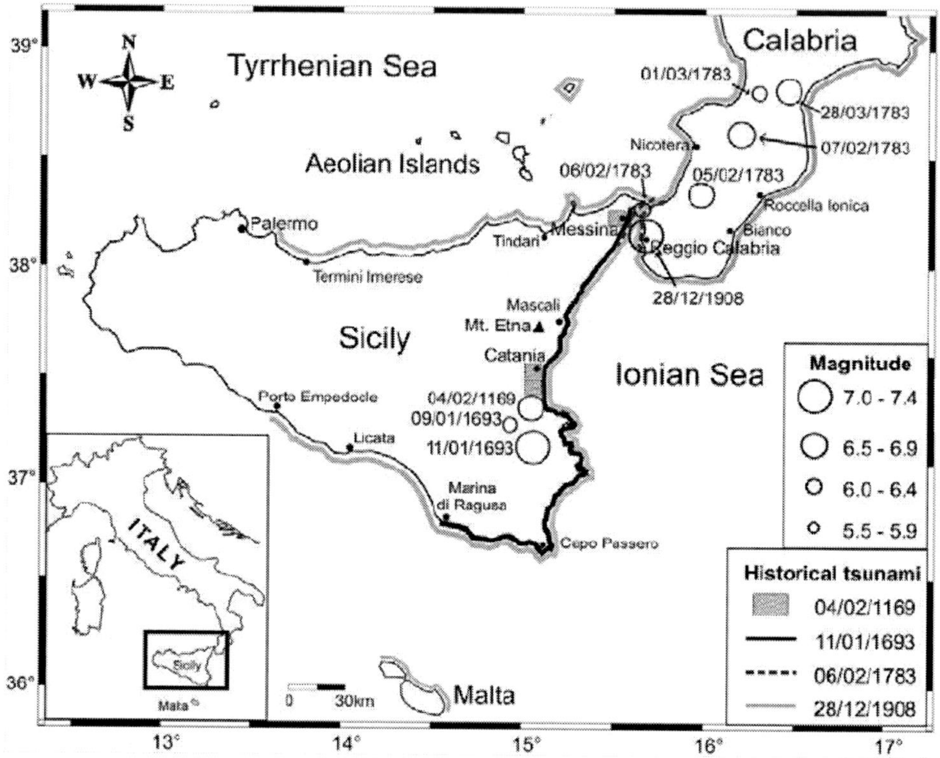

Figure 1. Epicentres (unfilled circles) of main earthquakes from the CPTI4 catalogue (Working Group CPTI, 2004). The known areas affected by main tsunamis are also shown.

2. Tsunami Deposits

In the past twenty years, deposits of several recent and historical tsunamis have been analyzed and described in detail (e.g., Clague et al., 1994; Sato et al., 1995; Dawson and Shi, 2000; Pinegina and Bourgeois, 2001; De Martini et al., 2003; Cochran et al., 2005 and 2006; Dawson and Steward, 2007, Pantosti et al., 2008). These observations have contributed to a general understanding of the characteristics of tsunami deposits, and to the establishment of criteria for their recognition. The source of tsunami sediments is primarily beach sand eroded from the shoreline and deposited on vegetated surfaces (Jaffe and Gelfenbaum, 2007) (Figure 2a). In some cases, a tsunami deposit can be represented by a single sheet-like layer of sand; elsewhere, it can be represented by chaotic sediment layers containing abundant stratigraphic evidence for sediment reworking and redeposition. The tsunami deposits commonly exhibit evidence of rapid deposition, such as gradation or massive structure and can be locally patchy and not diffused over the entire inundated surface (Clague and Bobrowsky, 1994). Tsunamis can also produce erosion, particularly in proximal or unvegetated areas.

To synthesize, sandy layers are interpreted to be tsunami deposits based on the following characteristics:

- The layers are found beyond storm-wave influence; lack of storm-wave influence is identified by changes in vegetation from beach grasses to less tolerant plants and by the presence of (sand-poor) peat; by distance from the shoreline or reconstructed shorelines, generally farther than 300 m (Morton et al., 2007). This is true for the ocean storms; in the Mediterranean, storms are substantially less powerful and their influence rarely exceed some tens of meters.
- The layers are similar to local beach sand, generally comprising well-sorted and well-rounded particles; and may include fragments of plants (Pinegina and Bourgeois, 2001).
- Layers are sheet-like, with typical thicknesses of a few millimetres to a few centimetres, generally thinning away from the shoreline (Pinegina and Bourgeois, 2001); the sediments occurring immediately below the sand layer exhibit evidence of wave erosion whilst in some places, plant leaves rooted in the underlying peat extend upward into and through the sand or on occasions are matted down at the base of the sand indicating burial by flooding.
- Macro- and microfauna contained in the layer indicate a marine provenance. The macro-faunal content can range from fish remains to a wide range of shell debris. Microfauna includes a wide species range of diatoms and foraminifera (e.g. Dawson et al., 1996), several of which originate in deep water. One of the major characteristics of diatoms and foraminifera found in tsunami deposits is the presence of a majority of broken individuals due to the turbulent water transport (Dawson and Shi, 2000).

However, given the large variability in the nature of tsunami sediments and of coastal environments, it is not surprising if tsunami deposits are not univocally identifiable, and other types of deposits may share some of the characteristics. Storm deposits most closely resemble tsunami deposits, although onshore storm deposits often have more complex layering than tsunami deposits (Figure 2b), do not extend inland more than ~100 m (Tuttle et al., 2004; Goff et al., 2004) and do not reach high elevations a.s.l. A further element of discrimination is the number of sandy layers in the sedimentary sequence: the storm deposits are generally more frequent than tsunamis thus, if the high energy layers were storm related they should be very numerous in the studied sequences (tens per 100 yrs). Also eolian deposits are sandy but they can be easily excluded because are typically very well sorted, made of very fine sand forming thicker, wedge-shaped layers; silt and very fine eolian sand are also disseminated in the peat. Flood deposits contains sand layers too but these are typically browner (contains organic matter) and muddier, and fluvial sediment are substantially less mature than those on the beach. Colluvium is poorly sorted, with angular grains (Pinegina and Bourgeois, 2001).

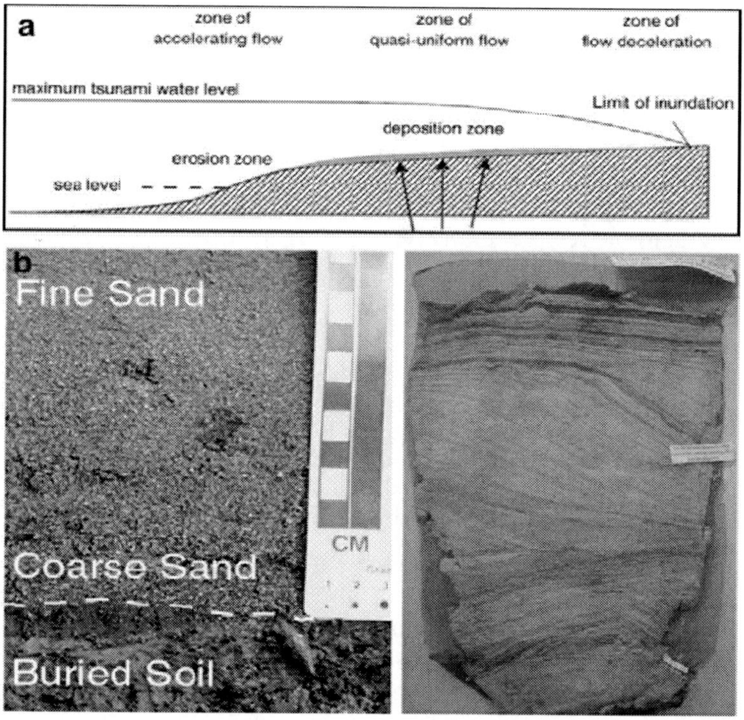

Figure 2. a) Conceptual model of tsunami sedimentation (from Jaffe and Gelfenbaum, 2007). b) On the left, deposit from 17 July 1998 tsunami in Papua New Guinea overlying brown muddy soil. The sandy tsunami deposit is fining upward and contains mud clasts ripped up from the soil (Morton et. al., 2007). On the right, storm deposits (Katrina 2005) form Outer Barrier Island, Mississippi (courtesy of J. Bourgeois).

3. Historical Tsunamis in Eastern Sicily

Because of the wealth of historical data available in Italy about the effects of seismicity of the past millennium, eastern Sicily and southern Calabria represent an exceptionally, favorable case for collecting, organizing and analyzing data on historical tsunamis. Recently, a database storing the effects of known tsunamis flooding the eastern Sicily and southern Calabria coasts has been compiled with the aim of reconstructing inundated areas and run-up distribution (Gerardi et al., 2008). In the following we summarize the historical data available for the main tsunamis (Table 1) with particular attention to the inundation extent.

3.1 ca. 1500 B.C. Santorini Tsunami

Investigations at Santorini indicate that the paroxysmal eruption (the Late Bronze Age or LBA eruption) of ca. 3500 BP was very large and may have generated multiple tsunamis. The LBA eruption had four phases, reflecting vent geometries and eruption mechanisms (Dominey-Howes, 2004 and reference therein), and tsunamigenic conditions prevailed during successive phases of the eruption. The exact date of the LBA eruption of Santorini has been the focus of an intense debate within the archaeological and scientific communities. Due to the lack of contemporary historical reports, evidence of the LBA tsunami comes from geological evidences onshore (Dominey-Howes, 2004; Bruins et al., 2008; Scheffers et al., 2008) and in the deep-sea floor of the Eastern Mediterranean (Kastens and Cita, 1981; Cita et al., 1996; Cita and Aloisi, 2000) testifying its regional importance.

3.2. The First Century A.D. Tsunami

Bonito (1691) and Mongitore (1743) report two earthquakes dated 14 and 33 A.D., respectively. The 33 A.D. earthquake is considered false because it coincided with the death of Christ, whereas for the 14 A.D. earthquake, both Bonito and (1691) Mongitore (1743), quoting Plinio, reported that the Tindari town was partly ruined and some towns were absorbed by the sea. This description is suggestive of the occurrence of a tsunami related to this earthquake. Due the uncertain dating for this historical period the 14 A.D. event could also coincide with the 17 A.D. earthquake that damaged many towns in Sicily and the area around Reggio Calabria (Guidoboni et al., 1994). The epicentral location of the 17 A.D.earthquake is roughly established in the northern side of Mt. Etna and its maximum intensity is VIII-IX MCS that translates to a M_{aw} = 5.14 (Working Group CPTI, 2004). No tsunami was described for the 17 A.D. event.

3.3. July 21, 365 Tsunami

According to Guidoboni et al. (1994) in his continuation of the Chronicon of Eusebiuos, written about 380 A.D. Jerome records: "There was an earthquake throughout the world, and the sea flowed over the shore, causing suffering to countless people in Sicily and many other islands". The source of this earthquake is located near Crete.

3.4. July 22, 963 Tsunami

G. Bucellino wrote: "An earthquake in Siria and Sicily, with destruction of towns, inundation of sea and thousands of casualties" (Mongitore, 1743). According to Guidoboni (1989) this event is false because it was impossible cross check this information in the Bucellino's paper.

3.5. February 4, 1169 Tsunami

The known inundated shoreline was between the Messina and Catania cities (Figure 1). In the Messina town the run-up overcame the city walls, inundating the streets (Fazello, 1560). Guidoboni and Comastri (2005) affirm that a contemporary chronicler, Roger of Hoveden reports that the sea receded from the shore in Catania, leaving many fishes behind on the sand to suddenly flow back. Effects of inundation were also described at the Simeto river mouth in the Catania plain (Sciuto Patti, 1896). Mongitore (1743) reports, quoting Tarcagnotta, that "the earthquake caused 25000 victims and many other were killed by the sea wave rising along the whole shore of the island" and citing Morigia, that "a river ran back fast, drowning more than 5000 people."

3.6. June 28, 1329 Tsunami

Many strong seismic shocks occurred after the beginning of an Etna eruption, in the western and eastern flanks of the volcano. Moreover, anomalous sea movements were reported in Mascali beach where some boats were carried into the sea by the waves (Fazello, 1560).

3.7. December 10, 1542 Tsunami

The earthquake, with an estimated magnitude of M_{aw} = 6.6 (Working Group, CPTI4), hit many localities in southeastern Sicily, producing damage in an area of about 6000 kmq (Boschi et al., 2000). Several localities (Sortino, Melilli, Lentini) were destroyed and the shock was felt in almost all parts of the island (Barbano and Rigano, 2001). After the earthquake, the town of Augusta was covered by the sea (Lacisio, 1543)

3.8. January 11, 1693 Tsunami

The M_{aw} = 7.4 (Working Group, CPTI4) 1693 earthquake caused destruction and heavy damage in most of eastern Sicily (Boschi et al., 1995). After the earthquake a large tsunami struck the whole eastern coast of Sicily between the Aeolian Islands and the old Port of Marina di Ragusa, Mazzarelli (Campis, 1694), in the southern Sicilian coast (Figure 1). The known length of inundated shoreline was of about 230 km (Gerardi et al., 2008).

The sea withdrew 60 steps (~100 m) and returned back overflowing the dock (Anonymous, 1693) at the Messina harbour. The largest inundation is described at Mascali, where the sea flooded the shore for about 1 mile inland (Boccone, 1697). The sea flooded Catania, inundating San Filippo square (now Mazzini square) and the farms around the town (Boccone, 1697). In the town of Augusta, the sea withdrew completely from the harbour and then violently came back over-passing the coast line of about 30 cubits (~165 m) (Bottone, 1718) and inundated the city as far as the San Domenico Monastery (Boccone, 1697).

3.9. February 5 and 6, 1783 Tsunamis

The 1783 seismic sequence was composed of five strong shocks occurred between February and March in southern Calabria (Figure 1). The earthquakes ruined many towns in Calabria and northeastern Sicily causing also important geomorphological changes (Boschi et al., 1995).

Tsunamis were observed after the two shocks of 5 February (M_{aw} = 6.9) and 6 Feb (M_{aw} = 5.9) (Working Group, CPTI4).

The 5 February tsunami hit the Ionian side of the Calabria coast from Bianco to Roccella Ionica (Graziani et al., 2006). The shore was flooded for about 1 mile inland at some locations (Galimi, 1783). On the Tyrrhenian coast small inundations at Scilla (Palestino in De Lorenzo, 1877), Punta del Pezzo (Sarconi, 1784), and two high sea waves at Nicotera (Minasi, 1785) were observed. The known length of inundated shoreline was about 40 km in the Ionian coast and 45 km in the Tyrrhenian one (Gerardi et al., 2008).

The 6 February tsunami affected the Sicilian coast from Messina to Torre Faro and the Calabria coast from Reggio Calabria to Scilla (Figure 1) for a total length of 40 km (Gerardi et al., 2008). The worst tsunami wave was about 16 m high and hit Scilla killing 1500 people. The sea water penetrated as far as 647 palms (~170 m) in the Livorno stream valley (Vivenzio, 1783). At Torre Faro, the tsunami wave flooded the shore for about 400 steps (~400 m) inland depositing a large amount of silt and dead fishes (Sarconi, 1784). In Messina, the sea penetrated inland as far as the fish market (Vivenzio, 1783).

3.10. December 28, 1908 Tsunami

The 1908 Messina earthquake (M_w = 7.1, Pino et al., 2000), is the most catastrophic event of the twentieth century in Italy; it produced extensive destruction over an area embracing southern Calabria and northeastern Sicily (Boschi et al., 2000 and references therein). The tsunami violently hit the southern Calabrian and the eastern Sicilian coasts causing further damage and causalities (Platania, 1909; Sabatini, 1910). The tsunami wave was also observed along the Tyrrhenian coast of the island and in the Sicily Channel as far as the Malta Islands (Figure 1, Platania, 1909; Baratta, 1910). The tsunami flooded up to 250 m in the north-eastern Sicilian coast between Scaletta and Acitrezza (Platania, 1909).

In Messina the tsunami inundated the harbour office and the St. Salvatore fortress, whereas at the Portalegni stream mouth the inundation penetrated 250 m inland (Platania, 1909). The waves flooded the town of Catania for more than 100 m inland depositing algae, posidonie, madrepore and millepore fragments, molluscs and many dead fishes; the shore was overflowed for about 700 m inland at the mouth of the Simeto river (Baratta, 1910). At Brucoli, about eight minutes after the shock, the sea withdrew for more than 200 m, then the sea flooded for about 50 m the field depositing boats, small fishes, echinoderms, crabs (Platania, 1909). In Augusta, around 20 minutes after the shock, inside the arbour a rumpus similar to that produced by the waves braking against rock-cliffs was felt; after few seconds, the sea flooded the beach with several waves, the greatest of which, the first one, rose to m. 1.75. Out of the harbour, the waves rose over 2 meters and propagated onshore for about 15 meters. The tsunami damaged the salt pan (Baratta, 1910). In Siracusa, after the downdraft, the wave rose about 2 m. Waters covered the banks and only some boats were damaged (Baratta, 1910). The known southernmost affected locality in Sicily was Capo Passero (Platania, 1909).

Table 1. Main known historical tsunamis in the eastern coast of Sicily

date	Seismic source	Hit localities/areas	Main sources
First century	unknown	Sicily	Bonito, 1695; Mongitore, 1743
365 July 21	Crete-Gortyna	Sicily and many other islands	Jerome 380
963 July 22	unknown	Siria and Sicily	G. Bucellino in Mongitore, 1743
1169 February 4	South-eastern Sicily	Messina, Catania, Sicily	Falcando, 12th cent.; Fazello, 1560; Mongitore, 1743
1329 June 29	Etna eruption and quake	Mascali (eastern Sicily)	Fazello, 1560
1542 December 12	South-eastern Sicily	Augusta	Lacisio, 1543
1693 January 11	South-eastern Sicily	Messina, Giardini, Mascali, Agnone, Augusta, Siracusa, Marina di Ragusa	AGdS, 1693; Burgos, 1693; Campis, 1694; Boccone, 1697
1783 February 5–6	Southern Calabria	Torre Faro, Messina, Scilla and Southern Calabria	Sarconi, 1784; Vivenzio, 1788; De Lorenzo, 1895
1908 December 28	Messina Straits	Eastern Sicilian coasts and southern Calabria	Platania, 1909; Baratta, 1910

4. Methodological Approach

On the basis of the historical reports, through the analysis of satellite images, aerial photographs and field surveys along the eastern Sicilian coast, five areas such as coastal lakes, marshes, lagoons and abandoned river channels were selected (Figure 3). These areas are potentially suitable for preserving tsunami deposits because they are characterized by coastal sedimentary traps, low-energy natural environment, continuous deposition, low erosion and accumulation of organic matter. These types of environment allow the recognition of the high-energy tsunami deposits that will appear completely anomalous and will contrast substantially with respect to those of low-energy commonly deposited.

In the selected areas, we explored several sites and carried out coring field surveys using both hand auger equipment and a vibracoring (with a gasoline-powered percussion hammer). Preliminary stratigraphical and sedimentological observations, together with photographs of the core deposits, and sampling for environmental and dating analyses, were performed directly in the field (Smedile et al., 2006; 2007). Once an interesting stratigraphic sequence was found, we were properly equipped to collect undisturbed samples (within specific pvc tubes), 100 cm long, 5 cm in diameter, down to 5 m total maximum depth. Coring was always accompanied by GPS surveys for the exact positioning of samples with respect to sea distance. A total of 64 test gouge cores were dug for a total length of 106 m.

Magnetic and X-ray analyses were performed on some selected cores sampled with PVC tubes to look for susceptibility variations and peculiar small-scale sedimentary structures (e.g., sharp contacts, convoluted layers, etc.). Then, cores were opened to perform accurate sedimentological descriptions and collection of all the needed samples for isotopic dating, paleontological, and electron microscope analyses (through a FE-SEM equipped with EDS). Macrofossils and foraminifera were analyzed at most sampled horizons. Samples have been submitted to washing with sieves of 63 μms, subsequently dried in oven, and analyzed with optic microscope. At least 100 foraminifera tests were counted from each sample where adequate concentration were present, whereas a qualitative analysis on macrofossils content was carried out. The main aim of all these analyses was to reconstruct the paleoenvironment of the site and to highlight possible anomalous deposits of marine provenance that can be interpreted as layers deposited by tsunami waves.

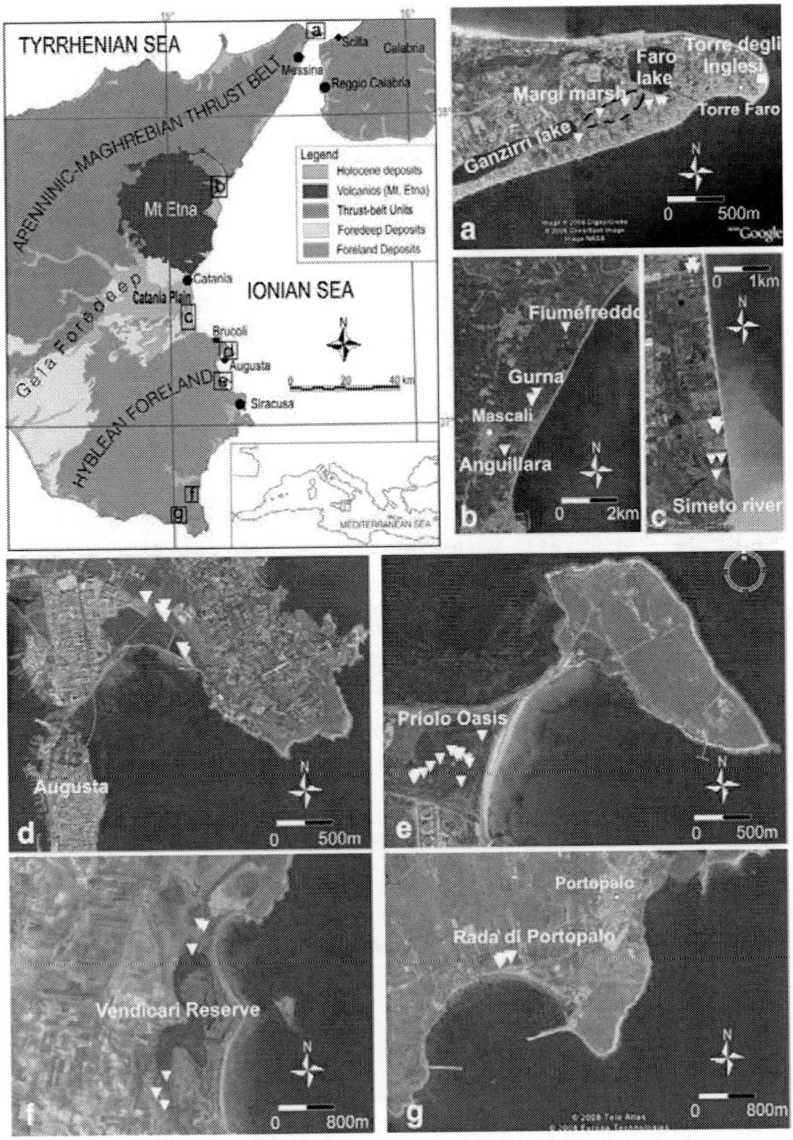

Figure 3. Main structural domains of SE Sicily; rectangles refer to the investigated areas shown to the right and below. Northeastern Sicily: a) Ganzirri area, the white square is the archaeological excavation at Torre degli Inglesi (Capo Peloro site) while the white triangles show the location of the gouge cores, hereinafter. Northeastern flank of Mt. Etna: b) the Fiumefreddo Oasis, Gurna and Anguillara sites; c) Catania plain. Gulf of Augusta area: d) Augusta site; e) Priolo site. Southeastern Sicily: f) Vendicari sites; g) Porto Palo site.

Finally, tephra characterization and radiocarbon analyses were used to constrain the age of the deposits, to derive sedimentation rates and to attempt correlation of the interpreted tsunami deposits with the historical events or to provide an age range of occurrence for the other events.

Radiocarbon dating was performed on charcoals, organic sediment and shells generally selected from the layers just above or below a suspicious layer. Occasionally, materials were selected from the layers themselves. We dated 11 samples (Table 2) through accelerator mass spectrometry (AMS) at the Poznan Radiocarbon Laboratory. Measured ages were dendrochronologically corrected according to Calib REV5.0.2 (Stuiver and Reimer, 2005).

Because we have a few radiocarbon datings from each site, some tsunami deposits are roughly constrained in age using average sedimentation rates. Each inferred tsunami deposit has been numbered from the top with the site label (e.g., GUR, ANG, etc.) and a progressive number.

If all the characteristics of tsunami deposits are fully satisfied (certain marine origin, high energy event, etc.) we attribute to the anomalous layers a high level of confidence; whereas if they show only some of the tsunami deposit characteristics or these are not certain (i.e., we cannot verify their origin) we assign a medium or low level of confidence (Table 3).

5. Study Areas and Results

The Ionian coastline of Sicily extends about 250 km from Capo Peloro to the north to Capo Passero to the south (Figure 3). The morphology of the coast and the local sedimentary environments (beach, dune, marshes, deltas, etc.) are strictly related to the geological setting of the island. In fact, to the north the coastline cross the south-verging thrust belt of the Peloritani mountains, and is characterized by fast uplift-rates ($\geq$ 1 mm/a) (Catalano and De Guidi, 2003), steep cliffs, narrow beach stretches and development of hydrographic basins drained by streams with short length (< 20 km). More to the south, the coastline develops onto the eastern flank of the Mount Etna, which is characterized by moderate uplift rate ($\leq$ 0.6 mm/yr, Di Stefano and Branca, 2002), basaltic lava flows, Holocene debris flow deposit and limited coarse-grained beaches. Southward, the eastward prograding deltaic system of the Simeto River (Catania alluvial plain) develops onto the tectonic depression of the Gela Foredeep, with the development of wide beaches, dunes and marshes. Finally, the southernmost part of the coast line develops onto the carbonatic sequences belonging to the moderate- to low-uplifted Hyblean foreland (Lentini et al., 1986) (Figure 3). In this area, several small lagoons developed along the coast.

Reconstruction of past environments on the coastal area of eastern Sicily shows changes from tidal inlet conditions to lagoons, dunes and alluvial deposits. We found peat environment only at one site (Fiumefreddo Natural Reserve), probably because of the local typical climatic conditions (temperature, precipitation, humidity, solar radiation, wind) as well as soil hydrology and chemistry.

Most of the sediments in the analyzed sequences represent steady-state deposition in relatively low energy regime coastal environments. However, in the dug cores we found also several anomalous units that may represent high-energy depositional events unexpected in a low energy environment.

In the following we describe, from north to south, the studied areas, the sampled sequences and the characteristics of the anomalous high-energy layers.

5.1. Northeastern Sicily

The Capo Peloro Peninsula, located in the Sicilian side of the Messina Strait, is characterized by the presence of two littoral lagoons, the Faro and the Ganzirri lakes; between the two lagoons we find a lowland area hosting the Margi marsh. This latter developed 4.6 ka ago parallel to the coastline with the progressive development of the dune which sealed off it from the sea (Antonioli et al., 2004). In this area off-shore bathymetry shows an abrupt deepening with a slope up to 20-25° for the first 200 m. The morphology of the coast, the low dynamics of the environment and the presence of a subsident area hosting the lakes and the marsh, are suitable for preserving tsunami deposits. Although the strong anthropization since Roman age (IV B.C.–IV A.D), we made some hand cores in the lowland area of the Margi marsh (Figure 3a).

The geological survey coupled with seven cores distributed in three sites (Figure 3a), revealed an old brackish lagoonal paleoenvironment evolving towards wetlands, in the uppermost part characterized by a coarse grained sedimentation. Although in the coastal plain of Ganzirri there is a suitable paleoenvironment for the preservation of tsunami deposits, no evidence of high energy events was found in the gouge cores.

However, in this area two anomalous clean, well sorted, sandy layers, interbedded in a organic colluvial sequence containing anthropogenic material, have been found at Torre degli Inglesi (TIA) archaeological excavation (Figure 3a) of Capo Peloro (Pantosti et al., 2008) and interpreted as tsunami deposits. The first anomalous layer (TIA-T2, Unit 90 in Pantosti et al., 2008) consists of a fine sand, made of inorganic components (sandstone remnants, quartz and micas), with sharp lower and upper contacts; it contains shell fragments (molluscs and corals), some planktonic and benthic foraminifera and algae remnants. The second and uppermost sandy layer (TIA-T1, Unit 1 in Pantosti et al., 2008) contains pebbles of decimetric size and shell fragments. The age of these layers has been constrained both with radiocarbon and archaeological dating. The age of the oldest layer is consistent with an earthquake occurred during the first century, probably the 17 A.D. earthquake that affected Reggio Calabria but for which no tsunamis was reported until now. The youngest layer is likely related to the 1783 February 6 earthquake (Pantosti et al., 2008) for which several report of tsunami exist.

Table 2. Measured and calibrated ages (according to Calib REV5.0.2 by Stuiver and Reimer, 2005) of the samples collected in the cores. ANG = Anguillara site, GUR = Gurna site, OSA = Augusta site, OPR = Priolo site. S is the core label; after the core label, the number indicate sample depth in cm. For the marine shells we adopted a ΔR = 124±77 yrs, to accommodate local reservoir effects

Sample	Type	Measured age B.P.	Calibrated age 2 σ
GUR-S10-403	Charcoal	3790 ± 30	2310–2135 B.C.
ANG-S01-220	Charcoal	335 ± 30	1455–1635 A.D.
ANG-S01-357	Charcoal	425 ± 30	1425–1510 A.D.
OSA-S06-96	Charcoal	1320 ± 30	650–770 A.D.
OSA-S06-158	Charcoal	1960 ± 30	40 B.C. –120 A.D.
OSA-S06-187	Organic sediment	2685± 30	900–800 B.C.
OSA-S06-193	Shell	3310 ± 30	1265–825 B.C.
OSA-S06-198	Organic sediment	2745 ± 30	975–820 B.C.
OPR-S11-43	Shell	890 ± 30	1420–1690 A.D.
OPR-S11-66	Shell	2460 ± 35	225 B.C.–220 A.D.
OPR-S01-158	Shell	3970± 35	2100–1635 B.C.

5.2. Northeastern Flank of Mt. Etna

In the northeastern flank of Mt. Etna area, three sites located behind an Holocene coastal dune have been analyzed: the Fiumefreddo Natural Reserve, Gurna and Anguillara (Figure 3b). These lowland sites were filled by clastic sediments derived from the erosion of the Peloritani Mts and the Etna lavas, and usually show high sedimentation rate related to the local fluvial environment. The presence of strong marine currents induces a distribution of the sediments parallel to the coastline. The emersion of sand bars above which the sand dunes start, causes the establishment of protected environments, suitable for the preservation of tsunami deposits.

The Gurna and Anguillara areas (Figure 3b) are characterized by two different types of paleoenvironment: in the first one a fresh-water shallow pond developed and persisted up to the present, while in the Anguillara site we found an alluvial environment, probably related to flow of the Macchia stream. Moreover, the Fiumefreddo Natural Reserve is characterized by a lacustrine environment.

5.2.1. Fiumefreddo Natural Reserve

This area has been investigated at one site (RIF) (Figure 3b), where we dug one core (RIF-S01) down to 2.5 m, located 500 m from the sea. The stratigraphic sequence appeared dominated by peat deposits. The presence of a 4 cm thick silty sand layer at about 130 cm depth (RIF-T1), characterized by rare planktonic foraminifera (*Orbulina universa*, *Globigerinoides* sp.) suggests a tsunami origin for this deposit (Smedile et al., 2006). Unfortunately, no dating is available yet.

5.2.2. Gurna Sites

The Gurna area has been investigated at two sites (Gurna north and south) (Figure 3b), where we dug four cores (GUR-S01/03 and GUR-S10) down to a maximum depth of 4.2 m, and at a max distance of 340 m from the sea. The stratigraphic sequence is composed mainly by fine sediments, silt and clay (Figure 4a), interrupted by four distinct coarse layers of sandy silt to coarse sand. These layers were found at depths of 42–48 cm (GUR-T1), 107–134 cm (GUR-T2), 170–178 cm (GUR-T3), made of quartz, volcanic clasts and with few vegetal remains in almost all the samples, and at 400–408 cm (GUR-T4) where a thin fine sand layer and some lens of sand with the same characteristics, were found. Morphoscopic analyses on clasts indicate that no primary Etna tephra deposits has been preserved in these sites. Unfortunately, all the samples collected for paleontological analysis are barren, nevertheless we tentatively interpret them as tsunami deposits.

Radiocarbon dating, performed on a charcoal collected within the deepest layer (GUR-T4) (GUR-S10-403, Table 2), constrains its age of deposition at 2310–2135 B.C., confining the occurrence of the four hypothesized tsunamis within the past 4000 yrs. Assuming limited erosion, the radiocarbon dating yields an average sedimentation rate of 0.8 mm/yr, excluding the tsunami deposits from the calculation, or 1.0 mm/yr, including them. Under these assumptions, the sedimentation rates can be used to constrain the age ranges for the other anomalous layers that result for GUR-T3 between 100–600 A.D.; for GUR-T2 between 650-1050 A.D. and for GUR-T1 between 1390–1780 A.D. Comparing the inferred ages of these tsunami deposits with the historical tsunami catalogue, we can tentatively associate the GUR-T3 tsunami layer to the 365 A.D. Crete tsunami and the youngest one to the 1693 tsunami. GUR-T2 could be the 1169 event taking into account the uncertainty of our age estimate and that for the 1169 event we have historical information about inundation along this coast.

5.2.3. Anguillara Site

The Anguillara area has been investigated at one site (ANG) (Figure 3b), where we dug three cores (ANG-S01/S03) down to a maximum depth of 5.2 m, at a distance of 280 m from the sea. The stratigraphic sequence is monotonous and mainly composed by fine sediments, dark grey clay and brown silt, with the exception of two well distinctive coarse layers made by fine sand to sand, at 60–70 cm (ANG-T1) and 195–200 cm (ANG-T2) depth (Figure 4b).

Morphoscophic analysis on several samples indicates that most of sediments contain a high amount of volcaniclastic material (scorias, lava clasts and crystals), nevertheless no primary fallout deposit has been preserved. All the samples collected for paleontological analysis are barren (quartz, volcanic clasts and few vegetal remains are predominant in almost all the samples) except for a very interesting detritic layer collected at 350–356 cm depth (ANG-T3), made by metamorphic rounded clasts in a dark grey silty-clay matrix, where two echinoderms fragments and high concentration of roots and seeds have been observed (Figure 4b). We interpret the two coarse layers and the abovementioned detritic deposit as sediments possibly left by tsunami waves.

Radiocarbon dating was performed on two charcoals sampled in the ANG-S01 core just below the deepest sandy layer at 195–200 cm and below the detritic layer at 350–356 cm depth, respectively (Table 2). The ages of samples ANG-S01-220 (1455–1635 A.D.) and ANG-S01-357 (1425–1510 A.D.) partially overlap suggesting a common local source for the two charcoals. If this is the case we are forced to use the age of the deepest sample as reference for our calculations.

Even though the uncertainties are large given the local environment, we can derive an average sedimentation rate from the sample ANG-S01-357 (1425–1510 A.D.) of 6–7 mm/yr for the past 580 yrs, excluding or including the tsunami deposits from the calculation, respectively.

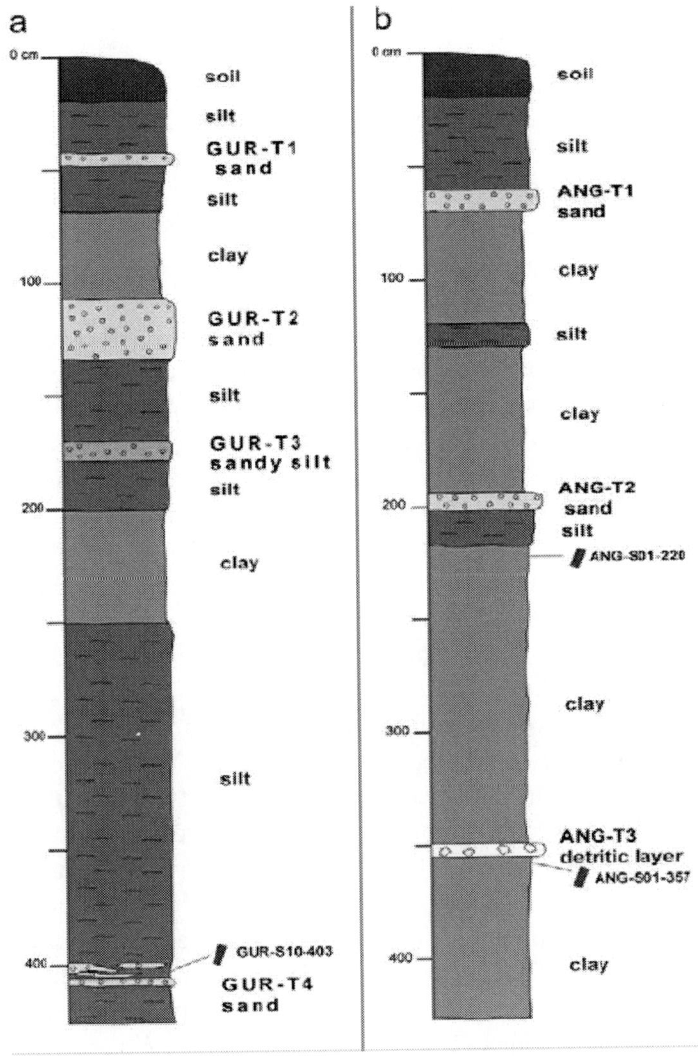

Figure 4. a) Gurna site: stratigraphic sequence of GUR-S10 log from 0 to 425 cm depth. GUR-T1, GUR-T2, etc. are anomalous layers. b) Anguillara site: stratigraphic sequence of ANG-S01 log from 0 to 425 cm depth. ANG-T1, ANG-T2, etc. are anomalous layers. The small black rectangles indicate depth of the dated charcoals (see Table 2).

On this basis, we can infer an age of deposition for the two anomalous layers: ANG-T3 just after 1425–1510 A.D., ANG-T2 in the range 1660–1710 A.D. and ANG-T1 in the range 1880–1920 A.D. Comparing these ages with the historical tsunamis, we can hypothesize that the ANG-T3, ANG-T2 and ANG-T1 anomalous layers found at the Anguillara site are associated to the 1542, 1693 and 1908 tsunamis, respectively.

5.3. Catania Plain

The Catania plain is the widest alluvial plain of Sicily (Figure 3), which extends from the southern flank of Mt. Etna to the northern boundary of the Hyblean Mts as part of the Quaternary Gela foredeep (Lentini et al., 1986). The drainage system, composed by a complex hydrographic network, is dominated by the Simeto, Dittaino, Gornalunga and San Leonardo rivers, which created the thick Holocene alluvial infill of the plain. This area is characterized by fluviatile environment and by subsidence, with some part of the ground surface sink below the sea level. It was selected following the historical records, that describe important inundation of the plain, and geomorphological observation that suggested the presence of possible trap-sites for tsunami deposits in an apparently marsh environment, e.g. the relict fluvial meanders to the north of the present Simeto delta and the abandoned channels of the Gornalunga river to the south of Simeto (Longhitano and Collella, 2007). Behind the anthropized littoral dune facing the Ionian Sea, we performed several cores (four sites and thirteen cores, Figure 3c) but, although the apparent condition of trap-sites, no suitable environment to preserve tsunami deposits was found: sedimentological and paleontological analyses of collected samples pointed out an extensive, thick alluvial and fluvial sedimentation showing a migration toward south of the coarser sediments from the Simeto river thought the time. Therefore, no anomalous layers were identified. It is interesting to notice that tephrostratigraphic investigation reveals that no primary tephra deposits are preserved despite the frequent historical explosive eruptions of Mt. Etna that has produced copious lapilli and ash fallout in this area. This latter information may confirm that the Catania Plain is not a suitable area for preservation of primary deposits due to high energy processes related to the Simeto River.

5.4. Gulf of Augusta Area

The Gulf of Augusta is located along the Ionian coast of the Hyblean mountain range, on the footwall of the Hyblean escarpment (Bianca et al., 1999). From a geomorphological point of view this area is characterized by high and low sectors tectonically controlled; in the low areas salt marshes are found which are particularly favourable for the deposition and preservation of the tsunami deposits. Two sites, characterized by the presence of salt marshes, have been surveyed: Augusta (Figure 3d) and Priolo (Figure 3e).

5.4.1. Augusta Site

In Augusta site (OSA), located near the local Hospital, we dug seven cores (OSA-S01/06 and OSA-S10) down to a maximum depth of 4.3 m, as far as 460 m from the sea (Figure 3d).

At this site we found one of the best evidence of paleotsunami deposit. In fact, the core OSA-S06 (Figure 5), a monotonous grey to dark grey fine silty deposit of brackish lagoonal environment, with abundant ostracods (*Cyprideis torosa*), whole gastropods (*Hydrobia* spp., *Pirenella conica*) and well preserved benthic foraminifera (*Ammonia parkinsoniana, A. tepida, Haynesina germanica*) (Figure 5b), is abruptly interrupted at a depth of about 190 cm by a yellowish bioclastic layer (OSA-T2) (Figure 5a). The latter is made of rare whole gastropods (*Hydrobia* spp., *Pirenella conica*), abundant shell fragments (mollusks, corals and echinoderms), rare ostracods, often broken benthic (*Ammonia* spp., *Bolivina* sp., *Cassidulina laevigata, Cibicides lobatulus, Haynesina germanica,* miliolids, *Pullenia bulloides, Rosalina* spp., etc.) and few badly preserved planktonic (*Globigerina* spp., *Globigerinoides* spp., *Globorotalia inflata, Turborotalita quinqueloba*) foraminifera (Figure 5c) (Smedile et al., 2007). This assemblage, supporting a clear marine origin for the bioclastic layer, coupled with the fact that it shows a sharp erosional basal contact and no evidence for layering or grading, allows us to interpret it as a tsunami deposit (De Martini et al., 2009). A similar fauna association has been detected in nearby core (OSA-S10) although the morphoscopic evidence is not so clear.

Furthermore, a gravel layer rich in vegetal remains (Figure 5a), angular calcarenites clasts and shell fragments, was detected at a depth of 162–173 cm (OSA-T1), it presents a sharp, most probably erosional, basal contact. However, no microfauna could be used to univocally establish the marine origin for this layer.

Radiocarbon dating was performed on 3 samples, collected just above, within and below OSA-T2 (Figure 5); OSA-S06-187, 193 and 198 (Table 2) gave coherent results, constraining the paleotsunami age to the interval 975–800 B.C.

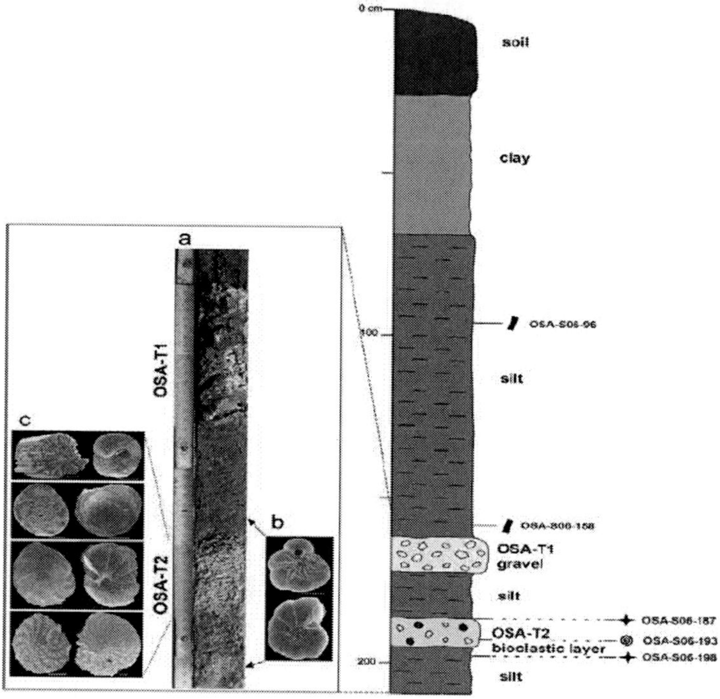

Figure 5. Augusta site: on the right stratigraphic sequence of OSA-S06 log from 0 to 210 cm depth. Small black rectangles, stars and the snail indicate depth of the dated charcoals, organic sediments and shell specimens, respectively (see Table 2). OSA-T1, OSA-T2 represent anomalous layers. On the left: a) picture of the OSA-S06 core between 140 cm and 210 cm showing the bioclastic layer (OSA-T2) and the gravel layer with angular calcarenites clasts (OSA-T1); b) SEM pictures (the scale bar represents 100 μm) of the benthic foraminifera assemblage typical of the lagoonal environment (from the top *Ammonia tepida* and *Haynesina germanica*) detected above and below the bioclastic layer; c) other SEM pictures (the scale bar represents 100 μm) of a coral fragment (photo on the uppermost left side), a marine planktonic foraminifera (*Globorotalia inflata* on the uppermost right side) and some selected benthic foraminifera species (clockwise starting from the uppermost second line *Rosalina bradyi, Asterigerinata mamilla, Lenticulina gibba, Cibicides lobatulus, Elphidium aculeatum, Elphidium crispum*) observed within the bioclastic layer.

We also performed radiocarbon dating on two charcoal fragments at 96 cm and 158 cm depth (OSA-S06-96 and OSA-S06-158 in Table 2), that yield calibrated ages of 650–770 A.D. and 40 B.C.–120 A.D., respectively. Combining the latter age with those obtained from OSA-S06-187 dating it is possible to constrain the age of the gravel deposit (OSA-T1) between 900 B.C. and 120 A.D.

Furthermore, radiocarbon dating suggests an average sedimentation rate of 0.7–0.5 mm/yr. These values are calculated taking into account (assuming erosion equal to tsunami deposits thickness) or excluding (assuming null erosion) the 19 cm thickness of the proposed tsunami deposits and the possibly artificial uppermost 35 cm, respectively. Considering these sedimentation rates, the age of the gravel deposit OSA-T1 can be narrowed to about 600–400 B.C. (De Martini et al., 2009).

5.4.2. Priolo Oasis Site

The Priolo area has been investigated at one site (OPR) (Figure 3e), where we dug seventeen cores (OPR-S01/17) down to a maximum depth of 4.2 m, as far as 530 m from the present coastline. In general, the stratigraphy is composed mainly by a monotonous sequence of fine sediments, from clay to silt. This is interrupted by four anomalous layers: two distinctive bioclastic layers (OPR-T1 at 35–40 and OPR-T2 at 50–60 cm depth), one detritic deposit at about 90 cm depth (OPR-T3) (Figure 6) and one sandy layer (OPR-T4, at about 160 cm depth). Combining micropaleontological and X-ray analyses we made the following observations: a) the entire stratigraphic sequence appears to belong to a lagoonal environment; b) both bioclastic layers, OPR-T1 and OPR-T2, are characterized by sharp basal contact, an abnormal concentration of shell fragments and whole gastropods (all arranged in a chaotic pattern) (Figure 6a), and shows an increment in the benthic foraminifera specific diversity; c) some of the benthic foraminifera recovered in both bioclastic layers comes from a shallow marine environment; d) the detritic deposit (OPR-T3, 2–3 cm thick) shows an anomalous assemblage made by macromammal bone fragments (Figure 6c) (personal communication of Prof. A. Kotsakis) together with rare and badly preserved benthic (*Cassidulina carinata*, *Cibicidoides pseudoungerianus*, *Melonis barleeanum*, *Planulina ariminensis*) and one planktonic (*Globigerinoides* sp.) foraminifera; e) in the 160 cm deep sandy layer (OPR-T4) marine microfauna appear well preserved, differently from the association characterizing the fine to very fine deposits above and below. Because of the marine provenance and fauna association and macroscopic evidence, we may interpret all the four layers as tsunami deposits (De Martini et al., 2009).

Dating of the deposits in the cores was performed both on the basis of tephra analysis and radiocarbon.

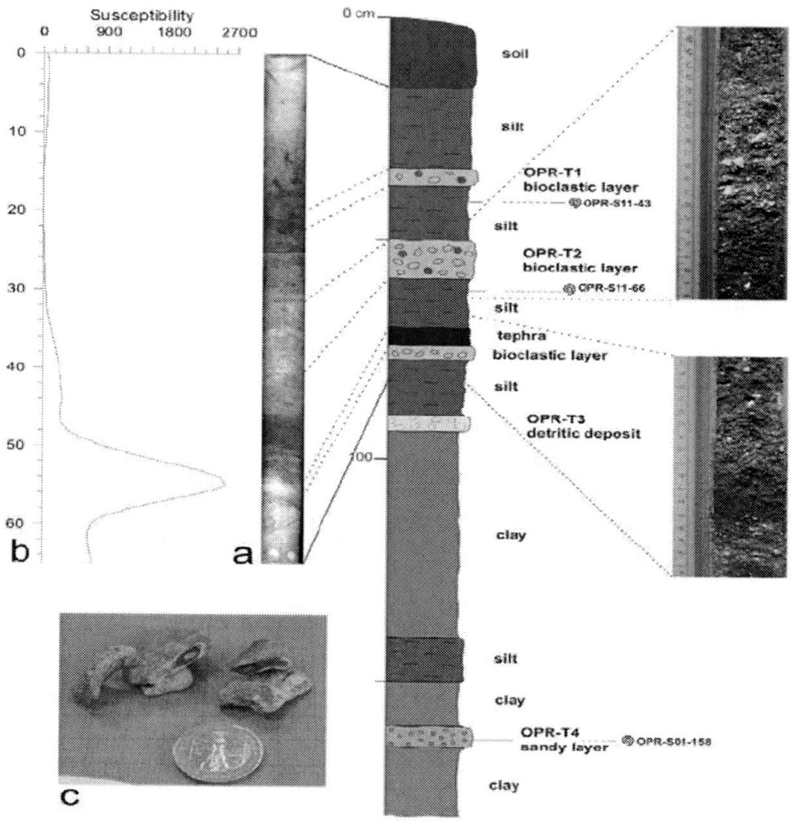

Figure 6. Priolo area: representative stratigraphy of OPR cores from 0 to 180 cm depth with the position of the dated shell samples, compared with a) X-Ray film of OPR-S11 core and b) magnetic susceptibility from the uppermost part of OPR-S11, please note how the bioclastic and volcanic layers show up on the film and the sudden increase of the susceptibility profile in coincidence of the volcanic layer; c) macromammal bone remnants found in the detritic deposit (OPR-T3). The pictures on the right show details of the bioclastic layer OPR-T2 (up) and the tephra layer (down).

Magnetic, petro-chemical and morphoscopic analyses on a black volcanic coarse sand, normally graded, found in six cores at about 70 cm of depth, allowed us to correlate it with the tephra from the 122 B.C. plinian eruption of Mt. Etna (Coltelli et al., 1998). In Figure 7, glass compositions of tephra recovered in Priolo site (De Martini et al., 2009) are compared with those of the 122 B.C. eruption. On the basis of the Total Alkali Silica classification diagram (TAS, Le

Maitre, 1989), they show alkaline affinity suggesting Etnean provenance and a similar chemical composition to the 122 B.C. products.

Radiocarbon dating was performed on three shell specimens (Table 2): these yielded an age of 1420–1690 A.D. at a depth of 43 cm (OPR-S11-43), 225 B.C.–220 A.D. at 66 cm (OPR-S11-66), and 2100–1635 B.C. at 158 cm (OPR-S01-158) (Figure 6).

The age of the shell OPR-S11-66 (225 B.C.–220 A.D.) is in good agreement with the 122 B.C. deposition age of the tephra layer found at 70 cm depth. Sample OPR-S1-9-158 constrains the age of the lowermost tsunami layer (OPR-T4 at 158-161 cm depth) at 2100–1635 B.C. The deposition of OPR-T3 occurred after 2100-1635 B.C and before the 122 B.C. (De Martini et al., 2009). The deposition of OPR-T2 occurred after the 122 B.C. Etna eruption and before 1420-1690 A.D. Finally, the radiocarbon dating performed at about 43 cm depth (OPR-S11-43) constrains the age of the youngest tsunami deposit (OPR-T1 at 35–40 cm depth) immediately after 1420–1690 A.D.

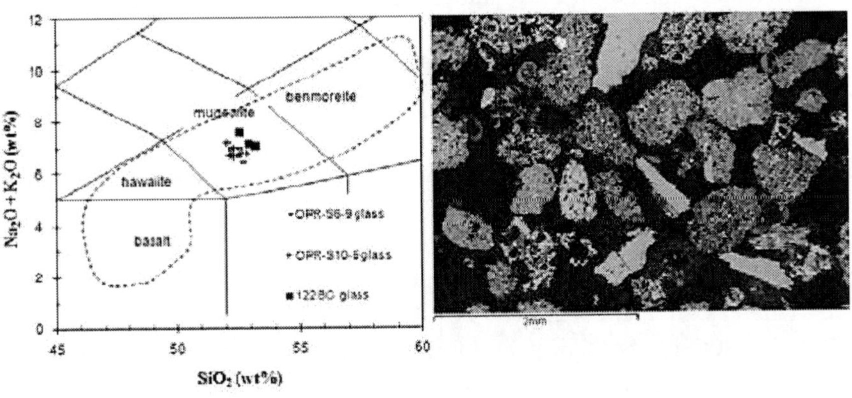

Figure 7. On the left side the Total Alkali Silica classification diagram (modified after De Martini et al., 2009) showing glass compositions of tephra recovered in Priolo site (OPR-S6 and OPR-S10 cores) and from the 122 BC plinian eruption of Etna (Coltelli et al., 1998). Dashed line includes the composition of Mt. Etna volcanics (Corsaro and Pompilio, 2004). On the right SEM picture of the OPR-S10-5 tephra sample composed of micro-pumices, loose crystals of plagioclase and minor lava lithics.

To attempt narrowing the age ranges of the tsunami layers OPR-T2 and OPR-T3 that are very wide, by combining tephrostratigraphic and C14 data in core OPR-S11 and S10 and the 122 B.C. tephra, that appear to be in good agreement, we estimated average sedimentation rates ranging from 0.25 mm/yr to 0.35 mm/yr for OPR-S11 and 0.33 mm/yr for OPR-S10 (De Martini et al., 2009). By using these rates, we can narrow the age of OPR-T2 between 160–320 A.D. and the age of OPR-T3 between 800–600 B.C. Comparing the event ages with the historical tsunamis, the layers at 35–40 cm (OPR-T1) can be associated to the 1693, whereas OPR-T2 at 50–60 cm depth could be the 365 or probably the 1169 event; no historical events seem to be related to OPR-T3 and OPR-T4, which could be paleotsunami events, although, considering the error in the age estimate, the last one could be correlate with the Santorini tsunami.

5.5. Southeastern Sicily

The SE Sicily area is located on the Hyblean foreland. There were found two areas with suitable morphologic characteristics to preserve tsunami deposits, the Vendicari Regional Natural Reserve (Figure 3f) and, more to the south, some ponds and salt marshes in Marzamemi and Porto Palo sites (Figure 3g). Both at Vendicari and Marzamemi/Portopalo a littoral dune separates lagoons from the coast; these restricted environment were also used as salt-pans during the past.

The paleoenvironmental analysis of the Vendicari Oasis shows a very recent tidal environment in the central part with an older brackish marine lagoonal behind, disturbed by the sedimentation of a local river in the southern part. We dug six cores down to a maximum depth of 2 m, at a maximum distance from the coastline of 800 m (Figure 3f). In the northernmost cores (VEN-S01/03), at a maximum distance from the coastline of 200 m, the stratigraphic sequence is composed by coarse sand to sand at the bottom and by clay to silty clay deposits above; whereas in the cores (ROV-S01/03) at a maximum distance from the coastline of 800 m the sedimentological analysis revealed a mostly silty clay sequence. No anomalous layers were identified.

The site tested in Porto Palo showed shallow ponds, predominately fresh water with marine inlet to the present day. This area was investigated at one site (RAP) (Figure 3g), where we dug three cores (RAP-S01/03) down to a maximum depth of 1.2 m, as far as 150 m from the sea. The stratigraphic sequence is composed by coarse sand to sand at the bottom and by clay to silty clay deposits above. Within the latter deposits, two thin and loose sandy beds with sharp lower contacts were found at about 8 and at 12 cm depth. Both layers present badly-preserved shallow marine benthic foraminifera with some planktonic foraminifera, similar to what was observed in all samples below. However, the deepest fine sand contains some well-preserved deep marine benthic foraminifera, sponge spicules, macrofossils and ostracods remnants, indicating a marine provenance and a possible association with a tsunami event. No data about rates of sedimentation at this site are available yet but, taking into account the very shallow depth of these sands, we have tentatively associated them with the 1908 tsunami.

6. CONCLUSION

The research of paleotsunami deposits in eastern Sicily is a new line of research only recently undertaken in Italy but also in the broad Mediterranean. Therefore, approaches to the identification and dating of tsunami deposits need severe testing, systematization and formalization. This work was carried out following a multi-theme approach and testing several different methodologies. Interesting and promising results came from an original combination of geomorphological, paleontological, X-ray, petro-chemical, morphoscopic and magnetic analyses.

In general, lagoonal environments and marsh areas are favourable sites to recognize tsunami deposits because of the presence of a peat or mud stratigraphy that allow preservation and easier identification of tsunami deposits than in the alluvial areas. In fact, in the alluvial areas, high-energy processes, erosion and intermittent deposition make the preservation of the thin tsunami deposits unlikely. Furthermore, the low sedimentation rate characterizing marsh environments provides, in few meters of sediments, the record of longer interval times (i.e., more tsunami events).

The tsunami deposit research in eastern Sicily showed evidence for clear tsunami inundations at the sites of Capo Peloro, Augusta and Priolo (Table 3). At Capo Peloro and Priolo sites, evidence of both historical and paleotsunamis were recognized; whereas at the Augusta site, we identified deposits associated with previously unknown inundations. Probable tsunami deposits were discovered both at Gurna and Anguillara sites. Three possible inundations in the past 580 yrs occurred at the Anguillara site, whereas four events in the past 4000 yrs were recognized at the Gurna site (Figure 8). No records of tsunami inundations in the Ganzirri area, in the Catania Plain or in the Vendicari Oasis have been found.

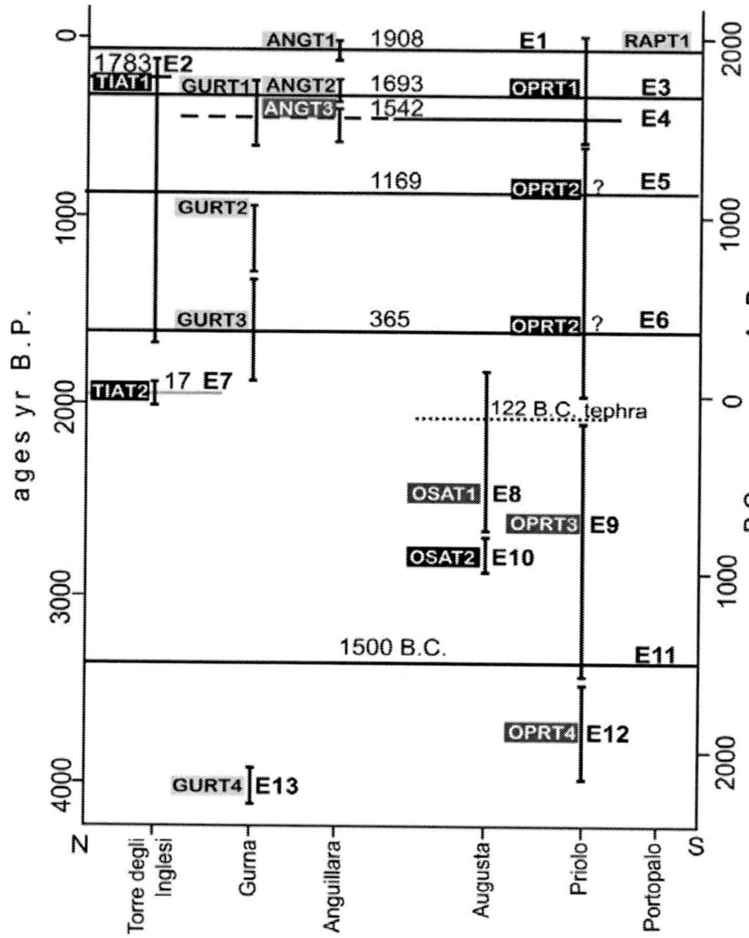

Figure 8. Space-temporal distribution of tsunami deposits identified in different sites along eastern Sicilian coast to be compared with historical tsunamis. The tsunami deposits are marked with a box and the core label: black boxes = high level of confidence (the layer has the most of the identification characters of tsunami deposit); medium grey boxes = medium level of confidence (the layer has many of the identification characters of the tsunami deposit); light grey boxes = low level of confidence (the layer has some of the identification characters of the tsunami deposit). E1, E2, ... E11 are the inundation events for which the relative deposits were found in eastern Sicily, marked with horizontal lines, whose length indicates extension of inundated coastline: black lines = historical tsunamis identified with their dates; grey lines = paleotsunami events. Dotted horizontal line marks the 122 B.C. tephra found in the Priolo site. Vertical lines indicate the chronological constraints for the event.

Table 3. Characteristics of the probable tsunami deposits found in eastern Sicily

Anomalous units (depth cm)	Layer characteristics	Macro and micro fauna and vegetal content	Confidence level
TIA-T1 (0–15)	Well sorted, silicoclastic sand layer with flat well-rounded cobbles	Shell fragments	High
TIA-T2 (125–130)	Grey clean silicoclastic well sorted sandy layer, with sharp, erosional contacts at base and top	Shell fragments of mollusks and corals, benthonic and planktonic foraminifera together with algae remnants	High
RIF-T1 (130–134)	Silty sand layer	Rare planktonic foraminifera	Low
GUR-T1 (42–48)	Well sorted fine sand with volcanic clasts; sharp basal contact	Vegetal remains; no macro-micro fauna	Low
GUR-T2 (107–134)	Well sorted fine sand	Vegetal remains; no macro-micro fauna	Low
GUR-T3 (170–178)	Fine sand, sharp basal contact	Vegetal remains; no macro-micro fauna	Low
GUR-T4 (400–408)	Very fine sand	Vegetal remains; no macro-micro fauna	Low
ANG-T1 (60–70)	Sand	Few vegetal remains; no macro-micro fauna	Low
ANG-T2 (195–200)	Fine sand	Few vegetal remains; no macro-micro fauna	Low
ANG-T3 (350–356)	Metamorphic rounded clasts in a dark grey silty-clay matrix	Echinoderms fragments and high concentration of roots and seeds	Medium
OSA-T1 (162–173)	Gravel, sharp basal contact	Vegetal remains; shell fragments, no micro fauna	Medium
OSA-T2 (187–195)	Bioclastic layer with a sharp erosional basal contact	Rare whole gastropods, abundant shell fragments (mollusks, corals and echinoderms), rare ostracods, often broken benthic and few planktonic foraminifera both from a marine environment; vegetal remains	High
OPR-T1 (10–15)	Bioclastic layer with a sharp basal contact	Abnormal concentration of shell fragments and whole gastropods (all arranged in a chaotic pattern), increasing in the benthic foraminifera specific diversity and few shallow marine benthic foraminifera species	High

Table 3. (Continued).

Anomalous units (depth cm)	Layer characteristics	Macro and micro fauna and vegetal content	Confidence level
OPR-T2 (30–40)	Bioclastic layer with a sharp basal contact	Abnormal concentration of shell fragments and whole gastropods (all arranged in a chaotic pattern), increasing in the benthic foraminifera specific diversity and recovery of few shallow marine benthic foraminifera species	High
OPR-T3 (90–)	Detritic deposit	Anomalous assemblage made by macromammal bone fragments together with rare and badly preserved benthic and one planktonic foraminifera	Medium
OPR-T4 (158–161)	Sand	Marine microfauna appears well preserved, differently from the association characterizing the fine deposits above and below	Medium
RAP-T1 (12–13)	Lens of fine sand with sharp basal contact	Some well preserved deep marine benthic foraminifera, sponge's spicules, macrofossils and ostracods remnants	Low

The deposits interpreted as tsunami-related rarely share the same characteristics (Table 3), which depend substantially on the depositional environment, from the local geomorphology and the type of the beach and offshore deposits. In general, the tsunami deposits represent layers different from the normal stratigraphic sequence, often coarse-grained chaotic units as a result of high-energy depositional events. When microfauna have been available, micropaleontological analysis confirmed the marine origin of these anomalous units. Dating of the tsunami units and their correlation with known historical tsunamis represent a critical issue for the final results of tsunami deposit studies. In this work, we have largely employed radiocarbon dating on charcoal fragments or shells and tephra analysis. The results obtained in this work are summarized in Figure 8.

Interestingly, we did not find evidence for the well-known 1908 tsunami, with exception of the Anguillara site and probably the Porto Palo site. This is likely because of the reconstruction activity after the 1908 Messina Strait earthquake and the intense twentieth-century urbanization of the costal areas that did not allow the preservation of the tsunami deposits. We identified the tsunami deposit associated with the 1783 event only at the Capo Peloro site, confirming historical information that confines the effects of this tsunami to the northern part of the Messina Strait. Moreover, in this site, we recognized an anomalous layer left by an inundation that occurred in the first century A.D., suggesting a possible similarity with the 1783 tsunami source.

We distinguished the deposit of the 1693 tsunami at Priolo site and probably also in Gurna site and Anguillara site, where also an anomalous layer probably associated with the 1542 tsunami was identified, but no evidence of older events have been discovered because of the high sedimentation rate. At Gurna, a tsunami deposit most likely related to the 1169 inundation was found; at Gurna and Priolo, records of inundation occurring during 100–600 and 220–300 A.D. could be related to the 365 tsunami, confirming that events with a faraway source could reach the coast of eastern Sicily.

We collected interesting evidence of paleoinundations occurring from 600–400 B.C. and 975–800 B.C. at Augusta, 800–600 B.C. at Priolo (De Martini et al., 2009), and, finally, between 2300 and 1635 B.C. at Gurna and Priolo sites (see Figure 8).

Despite the large uncertainties in ages, some tsunami deposits seem to be correlated with each other and/or with historical events. The number of paleoevents found with this study confirms that tsunami recurrence time in eastern Sicily is about 300–400 years.

By recognizing geological evidence of historical and paleotsunamis (useful in obtaining tsunami recurrence time, maximum inundation distance, elapsed time since the last tsunami event, etc.) through a multi-disciplinary study, we developed experimental approaches useful in identifying and dating tsunami deposits and, in addition, we provided a contribution for Civil Protection applications. The relevance of this type of study for Civil Protection is immediate in the field of tsunami scenario and modeling, and for time-dependent hazard assessment. This is especially true for eastern Sicily, which is a very densely inhabited coastland; for some areas, such as Augusta and Priolo, which have an important role as main national military and industrial sites; and for Capo Peloro, which has been chosen for building the bridge crossing the Messina Strait.

ACKNOWLEDGMENTS

This work was funded by the Italian Dipartimento della Protezione Civile in the frame of the 2004–2006 agreement with Istituto Nazionale di Geofisica e Vulcanologia – INGV (Seismological Project S2) and partially by the EU project TRANSFER. We are indebted to L. Gasperini from ISMAR-CNR-Geologia

Marina, Bologna for the help provided for the susceptibility and X-ray analyses. Many thanks are also given to E. Lipparini and G. Marozzi for the technical support kindly offered in the ISMAR-CNR labs. R. Azzaro, E. Baroux, M. Cosentino, G. D'Addezio, A. Mostaccio and some students helped with core logging and site survey. E. Gliozzi from the University of Roma Tre is thanked for assistance with ostracode and mollusk analyses. We are grateful to F. Cilea LIPU for the study of Priolo area; A. Cavallo, C. Freda and L. Miraglia for SEM analyses assistance; and M. Mancini for stratigraphic analysis. We also wish to thank J. Bourgeois for useful discussion on tsunami deposits.

REFERENCES

AGdS, (1693). Relazione dei luoghi che hanno sofferto in Sicilia nei Terremoti dal nove di gennaio fino al cinque di febbraio del 1693 che si sono sentite ventuno scosse, le prime tre grandi e le altre più lievi. In *Archivio General de Simancas, Secretaría de Estado, Negociación de Sicilia*, legajo 3507 (1693-1694). Consultas decretos y notas, n. 11, Palermo, IT.

Anonymous, (1693). Vera Relazione di quello che è successo nell'ultimo terremoto in Sicilia. Toulon, FR.

Baratta, M., (1910). La catastrofe sismica calabro-messinese (28 dicembre 1908), *Rel. Soc. Geogr. It.*, 1-496.

Barbano, M. S., Rigano, R., (2001). Earthquake sources and seismic hazard in Southeastern Sicily. *Ann. Geophys.-Italy*, 44 (4), 723-737.

Bianca, M., Monaco, C., Tortrici, L., Cernobori, L., (1999). Quaternary normal faulting in southeastern Sicily (Italy): a seismic source for the 1693 large earthquake. *Geophys. J. Int.*, 139, 370–394.

Boccone, P., (1697). *Intorno il terremoto della Sicilia seguito l'anno 1693;* Museo di Fisica, Venezia, IT, 1697; 1-31.

Bonito, M., (1691). Terra Tremante Overo Continuazione dé Terremoti Dalla Creazione del mondo fino al tempo presente. Napoli, IT, pp. 822.

Boschi, E., Ferrari, G., Gasperini, P., Guidoboni, E., Smriglio, G., Valensise, G., (1995). *Catalogo dei forti terremoti in Italia dal 461 a.C. al 1980*. ING-SGA, Bologna, IT, pp. 973 and CD-ROM.

Boschi, E., Guidoboni, E., Ferrari, G., Gasperini, P., Mariotti, D., Valensise, G., (2000). Catalogue of strong earthquakes in Itay from 461 B.C. to 1997. *Ann. Geophys. Italy*, 43 (4), 843-868 and CD-ROM.

Bottone, D., (1718). De immani Trinacriae terraemotu. Idea historico-physica, in qua non solum telluris concussiones transactae recensentur, sed novissimae anni 1717; Messina, IT.

Burgos, A., (1693). Distinta relazione avuta per lettera del P. Alessandro Burgos scritta ad un suo amico, che contiene le notizie fin'ora avute de' danni cagionati in Sicilia da terremoti a 9 e 11 gennaio 1693, Palermo-Roma, IT.

Bruins, H. J., MacGillivray, J. A., Synolakis, C. E., Beniamini, C., Keller, J., Kisch, H. J., Klügel, A., van der Plicht J., (2008). Geoarchaeological tsunami deposits at Palaikastro (Crete) and the Late Minoan IA eruption of Santorini. *J. Archaeol. Scie.*, 35, 191-212.

Campis, P., (1694). Disegno historico o siano l'abbozzate historiae della nobile e fidelissima città di Lipari (ms. 1694), Edited by G. Iacolino, Lipari, IT, 1991.

Caputo, M., Faita, G., (1984). Primo catalogo dei maremoti delle coste italiane. *Atti Accad. Naz. Lin.*, Roma, serie VIII, 17, 213–356.

Catalano, S., De Guidi, G., (2003). Late Quaternary uplift of northeastern Sicily: relation with the active normal faulting deformation. *J. Geodyn.*, 36 (4), 445-467.

Cita, M.B., Aloisi, G., (2000). Deep sea tsunami deposits triggered by the explosion of Santorini (3500 y BP), eastern Mediterranean. *Sediment. Geol.,* 135, 181–203.

Cita, M.B., Camerlenghi, A., Rimoldi, B., (1996). Deep-sea tsunami deposits in the eastern Mediterranean: new evidences and depositional models. *Sediment. Geol.,* 104, 155–173.

Clague, J. J., Bobrowsky, P. T., (1994). Evidence for a Large Earthquake and Tsunami 100–400 Years ago on Western Vancouver Island, British Columbia. *Quaternary Res.*, 41, 176–184.

Clague, J. J., Bobrowsky, P. T., Hamilton, T. S., (1994). A sand sheet deposited by the 1964 Alaska tsunami at Port Alberni, British Columbia. Estuarine. *Coastal and Shelf Science*, 38 (4), 413–421.

Cochran, U. A., Berryman, K. R., Mildenhall, D. C., Hayward, B. W., Southall, K., Hollis, C. J., (2005). Towards a record of Holocene tsunami and storms for northern Hawke's Bay, New Zealand. *New Zeal. J. Geol. Geop.*, 48, 507–515.

Cochran, U. A., Berryman, K. R., Zachariasen, J., Mildenhall, D. C., Hayward, B. W., Southall, K., Hollis, C. J., Barker, P., Wallace, L., Alloway, B., Wilson, K., (2006). Paleoecological insights into subduction zone earthquake occurrence, eastern North Island, New Zealand. *GSA Bulletin*, 118 (9/10), 1051–1074.

Coltelli, M., Del Carlo, P., Vezzoli, L., (1998). The discovery of a Plinian basaltic eruption of Roman age at Etna volcano (Italy). *Geology*, 26, 1095-1098.

Corsaro, R. A., Pompilio, M., (2004). Dynamics of magmas at Mount Etna. In *Mt. Etna: Volcano Laboratory*; Bonaccorso, A., Calvari, S., Coltelli, M., Del Negro, C., Falsaperla S.; Ed. AGU; Geophysical Monograph Series; 2004; Vol. 143, 91-110.

Dawson, S., Smith, D. E., Ruffman, A., Shi, S., (1996). The Diatom Biostratigraphy of Tsunami Sediments: Examples from Recent and Middle Holocene Events. *Phys. Chem. Earth*, 21(12), 87–92.

Dawson, A. G., Shi, S., (2000). Tsunami Deposits. *Pure Appl. Geophys*, 157, 875–897.

Dawson, A.G., Stewart, I., (2007). Tsunami deposits in the geological record, *Sediment. Geol.,* 200, 166–183.

De Lorenzo, A. M., (1877). Scilla inondata dal mare nella notte dopo il 5 febbraio 1783. In *Memorie da servire alla storia sacra e civile di Reggio e delle Calabrie, Cronache e Documenti rari*; Vol. 4, parte 4, fasc. 7; Reggio Emilia, IT, 1877; pp. 317-376.

De Lorenzo, A., (1895). *Un secondo manipolo di monografie e memorie reggine e calabresi*. Siena, IT.

De Martini, P. M., Pantosti, D., Barbano, M. S., Gerardi, F., Smedile, A., Azzaro, R., Del Carlo, P., (2006). Joint contribution of historical and geological data for tsunami hazard assessment in Gargano and eastern Sicily (Italy), *Abstracts of the Centennial Meeting of the Seismological Society of America*. Nr. 609

De Martini, P.M., Burratto, P., Pantosti, D., Maramai, A., Graziani, L., Abramson, H., (2003). Identification of tsunami deposits and liquefaction features in the Gargano area (Italy): paleosismological implication. *Ann. Geophys.-Italy*, 46 (5), 883-902.

De Martini, P. M., Barbano M.S., Smedile, A., Gerardi, F., Pantosti, D., Del Carlo, P., Pirrotta, C. (2009). A 4000 yrs long record of tsunami deposits along the Augusta Bay coast (Eastern Sicily, Italy): paleosismological implication. Submitted to Mar. Geol.

Di Stefano, A., Branca, S., (2002). Long-term uplift rate of the Etna volcano basement (southern Italy) based on biochronological data from Pleistocene sediments. *Terra Nova*, 14, 61–68.

Dominey-Howes, D.T.M., 2004. A re-analysis of the Late Bronze Age eruption and tsunami of Santorini, Greece, and the implications for the volcano-tsunami hazard. *J. Volcan. Geotherm. Res.,* 130, 107-132.

Falcando, U., 12th cent. La istoria o liber de Regno Sicilie. In *Fonti per la storia d'Italia*, Siragusa, G.B.; SS; Rome, IT, 1897; Vol. 22.

Fazello, T., (1560). *De rebus Siculis decades duae*; Palermo, IT.

Galimi, P., (1783). Lettera di Procopio Galimi al signor D. Giuseppe Vairo sui tremuoti di Calabria dell'anno 1783, s.l.

Gerardi, F., Barbano, M. S., De Martini, P. M., Pantosti, D., (2008). Discrimination of Tsunami Sources (Earthquake versus Landslide) on the Basis of Historical Data in Eastern Sicily and Southern Calabria. *Bull. Seismol. Soc. Am,* 98 (6), 2795–2805, doi: 10.1785/0120070192.

Gianfreda, F., Mastronuzzi, G., Sansò, P., (2001). Impact of historical tsunamis on a sandy coastal barrier: an example from the northern Gargano coast, southern Italy. *Nat. Hazard Earth Sys.*, 1, 213–219.

Graziani, L., Maramai, A., Tinti S., (2006). A revision of the 1783–1784 Calabrian (southern Italy) tsunamis. *Nat. Hazard Earth Sys*., 6, 1053–1060.

Guidoboni, E., (1989). *I terremoti prima del Mille in Italia e nell'area mediterranea*. Storia archeologia sismologia, Bologna, IT, 768 pp.

Guidoboni, E., Comastri, A., Traina G., (1994). *Catalogue of ancient earthquakes in the Mediterranean area up to the 10th century*; Istituto Nazionale di Geofisica, Rome, IT, 504 pp.

Guidoboni, E., Comastri, A., (2005). *Catalogue of earthquakes and tsunamis in the Mediterranean area from the 11th to the 15th century*. INGV, Rome, IT, 1037 pp

Jaffe, B. E., Gelfenbuam, G., (2007). A simple model for calculating tsunami flow speed from tsunami deposits. *Sediment. Geol.*, 200 (3-4), 347–361.

Jerome, (380). *Chonicon Eusebii, Eusebius Chonicon*, ed R. Helm, GSC 47, Berlin 1956.

Kastens, K.A., Cita, M.B., (1981). Tsunami induced sediment transport in the abyssal Mediterranean Sea. *Geol. Soc. Am. Bull.,* 89, 591–604.

Lacisio, P., (1543). Lettera di Paolo Lacisio ad Amerbach, Strasburgo 7 marzo 1543. In *Die Amerbachkorrespondens*, Hartmann A, Basel, SW, 1858, Vol. 5, pp. 414.

Le Maitre RW (ed.) (1989). A classification of igneous rocks and glossary of terms. Blackwell Scient Publ, Oxford. 193 pp.

Lentini F., Carbone S., Cugno G., Grasso M., Scamarda G., Sciuto F., Montanari L., Romeo M., Ferrara V., 1986. Geological map of the north-eastern Hyblean sector. Ed. S.E.L.C.A. Map scale 1:50.000, Firenze, IT.

Longhitano, S., Coltella, A., (2007). Geomorphology, sedimentology and recent evolution of the anthropogenically modified Simeto River delta system (eastern Sicily, Italy). *Sediment. Geol.,* 194, 195–221.

Mastronuzzi, G., Sansò, P., (2000). Boulders transport by catastrophic waves along the Ionian coast of Apulia (Southern Italy). *Mar. Geol.*, 170, 93–103.

Mastronuzzi, G., Sansò, P., (2004). Large boulder accumulations by extreme waves along the Adriatic coast of southern Apulia (Italy). *Quatern. Int.*, 120, 173-184.

Mastronuzzi, G., Pignatelli, C., Sansò, P., Selleri, G., (2007). Boulder accumulations produced by the 20th of February, 1743 tsunami along the coast of southeastern Salento (Apulia region, Italy). *Mar. Geol.*, 242, 191–205.

Minasi, G., (1785). Continuazione ed appendice sopra i tremuoti descritti nella relazione colla data di Scilla de 30 settembre 1783 con altro che accadde in progresso. Messina, IT.

Mongitore, A., (1743). Istoria cronologica de' terremoti di Sicilia. In *Della Sicilia ricercata nelle cose più memorabili*, Palermo, IT, Vol. 2, pp. 345-445.

Morton, R. A., Gelfenbuam, G., Jaffe, B. E., (2007). Physical criteria for distinguishing sandy tsunami and storm deposits using modern examples. *Sediment. Geol.*, 200 (3-4), 184-207.

Pantosti, D., Berryman, K., Yeats, R.S., Kinusaga, Y., (2003). Preface, Ten years of Paleoseismology in the ILP. Progress and prospective. *Ann. Geophys.-Italy*, 46 (5), V-VII.

Pantosti, D., Barbano, M.S., Smedile, A., De Martini, P.M., Tigano, G., (2008). Geological Evidence of Paleotsunamis at Torre degli Inglesi (northeast Sicily). *Geophys. Res. Lett.*, 35, L05311doi:10.1029/2007GL032935.

Pinegina, T. K., Bourgeois, J., (2001). Historical and paleotsunami deposits on Kamchatka, Russia: long-term chronologies and long-distance correlations. *Nat. Hazard Earth Sys.*, 1, 177–185.

Pino, N.A., Giardini, D., Boschi, E., (2000). The December 28, 1908, Messina Straits, southern Italy, earthquake: Waveform modelling of regional seismograms. *J. Geophys. Res.*, 105, NO B11, 25473-25492.

Platania, G., (1909). Il Maremoto dello Stretto di Messina del 28 dicembre 1908. *Boll. Soc. Sism. Ital.*, 13 (1908-1909), pp. 369-458.

Ruiz, F., Rodríguez-Ramírez, A., Cáceres, L. M., Rodríguez Vidal, J., Carretero, M.I., Abad, M., Olías, M., Pozo, M., (2005). Evidence of high-energy events in the geological record: Mid-holocene evolution of the southwestern Doñana National Park (SW Spain). *Palaeogeogr. Palaeocl.*, 229 (3), 212-229.

Sabatini, V., (1910). *Contributo allo studio dei terremoti calabresi*; Tip. Bertero; Roma, IT.

Sarconi, M., (1784). *Istoria de' fenomeni del tremoto avvenuto nelle Calabrie, e nel Valdemone nell'anno 1783*, posta in luce dalla Reale Accademia delle Scienze, e delle Belle Lettere di Napoli, 2-4.

Sato, H., Shimamoto, T., Tsutsumi, A., Kawamoto, E., (1995). Onshore tsunami deposits caused by the 1993 Southwest Hokkaido and 1983 Japan Sea earthquakes. *Pure Appl. Geophys.*, 144 (3/4), 693–717.

Scheffers, A., Kelletat, D., Vött, A., Matthias May, S., Scheffers, S., (2008). Late Holocene tsunami traces on the western and southern coastlines of the Peloponnesus (Greece). *Earth Planet. Sc. Lett.*, 269, 271–279.

Scicchitano, G., Monaco, C., Tortorici, L., (2007). Large boulder deposits by tsunami waves along the Ionian coast of south-eastern Sicily (Italy). *Mar. Geol.*, 238, 75-91, doi: 10.1016/j.

Sciuto Patti, C., (1896). Contributo allo studio dei terremoti in Sicilia. *Atti Accad. Gioenia Scie. Nat.*; Catania, IT; S. IV, 73, 9, 1-34.

Smedile, A., Barbano, M.S., De Martini, P.M., Gerardi, F., Pantosti, D., Pirrotta, C., Azzaro, R., Cosentino, M., D'Addezio, G., Del Carlo, P., Guarnieri P., (2006). Multidisciplinary study to identify tsunami deposits in eastern Sicily, *XXV Convegno GNGTS. Riassunti estesi delle comunicazioni*, Roma, 28-30 Novembre 2006, 325-328.

Smedile, A., Barbano, M.S., De Martini, P.M., Pantosti, D., Del Carlo, P., Gerardi, F., Guarnieri, P., Pirrotta, C., Cosentino, M., (2007). In search of tsunami deposits along the eastern coast of Sicily (Italy), *IUGG XXIV 2007*, Perugia, July 2-13.

Stuiver, M., Reimer, P.J., (2005). Radiocarbon calibration program CALIB REV5.0.2, copyright, available at http://calib.cub.ac.uk/calib/

Tinti, S., Maramai, A., Graziani, L., (2001). A new version of the European tsunami catalogue: updating and revision. *Nat. Hazard Earth Sys.*, 1, 255–262.

Tinti, S., Maramai, A., Graziani, L., (2004). The new catalogue of Italian tsunamis. *Nat. Hazards*, 33, 439-465.

Vivenzio, G., (1783). Istoria e teoria dè tremuoti in generale ed in particolare di quelli della Calabria, e di Messina del MDCCLXXXIII. Napoli.

Working group CPTI, (2004). *Catalogo Parametrico dei Terremoti Italiani*, version 2004 (CPTI04). INGV, Bologna. http://emidius.mi.ingv.it/CPTI/.

In: Recent Progress on Earthquake Geology ISBN: 978-1-60876-147-0
Editor: Pierpaolo Guarnieri, pp. 147-163

Chapter 6

EARTHQUAKES AND SHALLOW STRUCTURES IN SOUTH ADRIA: EVIDENCE OF RECENT INVERSION TECTONICS

Del Ben Anna
University of Trieste, DiSGAM, Via Weiss, 1 34127 Trieste – Italy

ABSTRACT

The Adria foreland flexure below the Calabrian and Hellenic Chains generated a system of normal faults, mainly NNW-SSE oriented, during the Pliocene-Pleistocene age. These are clearly evident on seismic reflection profiles, and are generally interpreted as the present main tectonic features of the foreland. A detailed analysis of the available data have instead highlighted, along the western margin and in the South Salento offshore, the presence of several recent compressive features which have been assigned to inversion tectonics, probably developing since the Middle Pleistocene age. The present regime would be the result of compressive/transpressive stress domain, as testified by several earthquakes, transferred to the deep Adria crust by the Calabrian and Hellenic Chains. The fault blocks, previously tilted during the Upper Pliocene/Lower Pleistocene age, have successively back rotated by inversion tectonics, so to produce the uplift of the rigid blocks, the seismic events recorded in the region and the compressive deformation, especially evident inner the upper soft sediments. In this way, the morphology of the seismic area of the South Salento offshore is characterized by gentle surface swells, typically placed above the hanging wall of the post-Miocene extensive faults. The present SSE-ward migration of the Calabrian Arc, nearly parallel to the western Apulia margin, transferred also the important strike-slip component to the deep level of the foreland.

INTRODUCTION

In the Puglia offshore the Apulia carbonate top deepens east-ward below the Albanides-Hellenides and west-ward below the Calabrian thrust belts. The carbonate platform started taking shape in the Late Triassic above a Paleozoic-Triassic continental sequence which is thicker than 7000 m, as Patacca et al. (2008) who interpreted the data in correspondence to the eastern termination of the CROP 11 seismic profile, near the Central Adriatic Italian coast. This sequence covers the pre-Cambrian crystalline basement. On the Adria plate, during the Early Jurassic age, the Ionian/South Adriatic-Umbria Marche pelagic basin started developing (Aubouin et al., 1970; Mattavelli et al., 1991), adjacent to the shallow water domain of Apulia platform (figure 1), that persisted until the Cenozoic age. In the South Adriatic the margin is represented by a NW-SE slope which E-prolongation is outcropping and deformed in the Albanian onshore (Nicolai and Gambini, 2007). Later clastic sediments of the Albanides-Hellenides Chain covered the Ionian/South Adriatic Basin and the eastern margin of the carbonate platform. On the west side the platform turned into the Ionian oceanic basin, currently in part subducted underneath the South Apennine/Calabrian Chain, as the deep inclined reflectors in the CROP-M profiles acquired in the Tyrrhenian Sea clearly represent (Finetti, 2005). The associated foredeep basin developed while migrating E-ward on the Apulia platform, here also covered by clastic sediments. The opposite chains brought about remarkable stress regime on the common foreland: Tozzi et al. (1988) highlights a sharp change from the older (Upper Cretaceous) anticlockwise, to the younger (Cenozoic) clockwise rotation of the Adria foreland, reaching up to 20-25° for the South Salento. Gambini and Tozzi (1996) identified the South Salento-North Kerkira Fault Zone, an E-W lineament separating the block of the Salento Peninsula by that one of the Apulian Plateau.

Argnani et al. (2001) evidenced a normal faults system cutting the base of the Plio-Quaternary sedimentary sequence in the southern offshore of the Salento peninsula. In their interpretation of some reflection profiles they define some grabens NW-SE oriented, partially insisting on the area of high seismic risk (Castello et al., 2006) and which are the prolongation of the grabens existing in the Salento onshore. A strong historical earthquake in 1743 damaged mainly the

Salento peninsula and the Ionian Isles on the western Greek offshore (160 casualties in Puglia and 100 in Greece, M=6.9, Catalogo CPTI04). Armigliato et al. (2007) explain this event as a complicate focal mechanism probably connected to a slump. Tinti et al. (2004) record evidence of tsunami in the port of Brindisi and Mastronuzzi et al. (2006) associate it to some "boulders" along the Apulia shore.

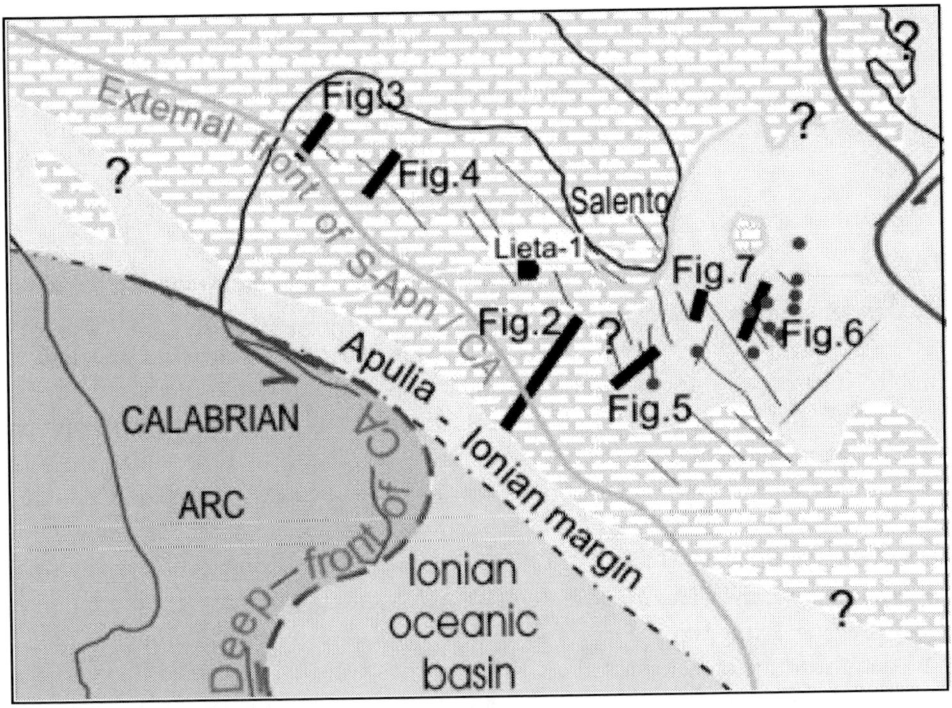

Figure 1. Map of the studied area with position of the seismic reflection profiles shown in this chapter. The external and the deep crustal front of the South-Apennine/Calabrian Arc (CA), the outline of the normal fault system (thin black lines) on the Adria foreland and the Apulia carbonate platform domain (brick symbol) are represented. The area of the South Adria plate represents the Upper Triassic/Lower Jurassic platform covered by Apulia main carbonate platform and other isolated platforms pelagic sediments during the Middle Jurassic/Cenozoic age. The seismic events in the period 1981-2002 by Castello et al. (2006) are indicated by dots.

During the last 20 years, the earthquakes occurring near the seismic risk area of the Salento offshore, reach a magnitude of 3-4 degree in the Richter scale and their ipocentral depth ranges between 30-40 km (Castello et al., 2006). D'Ingeo et al. (1980), Favali et al. (1990) and Vannucci et al. (2004) determined the focal mechanisms of main events, at a depth ranging between 0-50 km, implying a main compressive component (mainly directed NE-SW) with a secondary (left) strike-slip component. D'Ingeo at al. (1980) hypothesis claim the presence of a plate margin dividing the Adriatic microplate from the African plate located across the Otranto Channel. Argnani et al. (2001) suggested that the seismicity of this area is provoked by the high curvature of the Adria plate which is due to its underthrusting below the Calabrian and Hellenic Chains. These authors assert that there are two opposite tectonic regimes: extensive in the outer arc and compressive in the inner arc, separated by a neutral surface.

SEISMIC DATASET

The seismic reflection profiles interpreted in this chapter are mainly based on old datasets acquired for hydrocarbon exploration of the peri-Italian seas since the '60's (Mattavelli and Novelli, 1990). The seismic dataset of Zones D and F (public dataset, source the Italian Ministry of Industry) was acquired in the offshore of the Ionian and South Adriatic Italian coast with medium resolution and are often calibrated by several boreholes (Mattavelli and Novelli, 1990). In 1971, crustal MCS profiles were carried out by Osservatorio Geofisico Sperimentale (OGS) of Trieste, sponsored by the Italian National Council of Research (CNR). In 1991 the R/V OGS Explora collected some deep crustal MCS profiles in the Calabrian Arc/Salento offshore, within the framework of the Italian Deep Crustal Exploration Project (CROP) (Scrocca et al., 2003; Finetti, 2005). The acquisition parameters of the CROP profiles were designed for deep crustal exploration, but they show to be useful also to achieve information concerning the shallow Plio-Quaternary sequence, specifically analyzed in this study.

SEISMIC INTERPRETATION

The interpretation of public seismic profiles suggests that, during the Cenozoic age, the thick carbonate platform of the Adria plate, in the offshore of the South Salento peninsula, was SE-ward bordered by a complex margin: a reef is described along the Castro/S. Maria di Leuca shore, developed until the Messinian age, as recognized along the SE-Adria coast by Bosellini et al. (2001). The offshore prolongation of this reef is clearly recognizable along some seismic profiles across the Otranto Channel (Del Ben et al., 2008b). On the South a complex alternation of isolated platforms and pelagic domains seems to characterize the southern Adria plate (Del Ben et al., 2008b). This articulated structural condition generally complicate the seismic interpretation of the pre-Pliocene sedimentary sequence, which is often disturbed by the noise due to multiple reflectors, diffractions, scattering, etc. Anyway the base of the Plio-Quaternary sequence, characterized by an unconformity with high reflectivity, is always clearly interpretable.

The seismic profiles show the tilting of the Adria foreland, W-ward below the Calabrian front and E-ward below the Hellenic front. As already described by many authors (Royden et al., 1987; Royden, 1988; de Alteriis and Aiello, 1993; Argnani et al., 2001), the quasi-symmetric curvature of the plate, due to the loading of the two opposite chains, generated normal fault systems with a main NW-SE direction. Some of these faults, which are evident in figure 2a, separate the platform in tilted blocks below and towards the Calabrian front.

The identification of the precise dating of these faults along the west margin of the plate, is not simple because of the lack of valid calibrations on the region, but we can consistently assert they are post-Messinian, during the Ionian subduction below and the Adria plate tilting toward the front of the chain. Moreover they are likely to be mainly activated before the oblique collision of the Calabrian Arc with the continental Adria plate. The collision determined a slower migration of the chain, probably connected to the Early Pleistocene swift decrease of extensive tectonics in the Marsili Basin, after its fast opening (Nicolosi et al., 2006).

Nevertheless the main evidence of this study is represented by some compressive structures interpreted on the base of the fault-related folding theories (Shaw et al., 2005). These structures involve the Plio-Quaternary sequence of the Adria foreland and are generally very gentle, revealing a diffuse, recent compressive deformation of small value. Evidence is recognizable in the Taranto Gulf (figure 3) where the proximity of the Calabrian front makes reasonable an external expression of the compressive regime in the near outer region. This deformation develops mainly through the favourable detachment surface represented by the top of the Apulia platform, which is W-ward tilted and covered by onlapping Plio-Quaternary sediments. But other decollements inside this last thickness also allow the formation of fold, thrust and back-thrust.

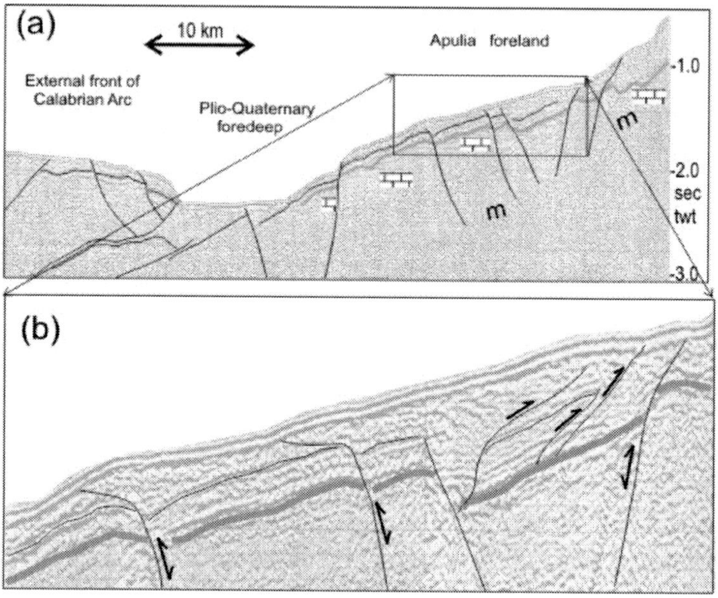

Figure 2. (a) Seismic profile MS-26 (part) showing the tilting of the Adria plate below the front of the Calabrian Arc (vertical exaggeration at the sea bottom is about 13x): this rotation determined fracturing of the plate through a normal fault system. “m” is the sea bottom multiple, the black line corresponds to the base of Plio-Quaternary sequence, the grey line is the top of the Apulia carbonate platform. Note the important back-tilting phenomenon connected to the NE-verging faults. (b) Detail of the MS-26 seismic profile, where the sea bottom and the shallow reflectors show gentle deformation. This last is interpreted as effect of the recent compressive tectonics, often inverting previous normal faults, specially those ones, more sloping, that are E-verging.

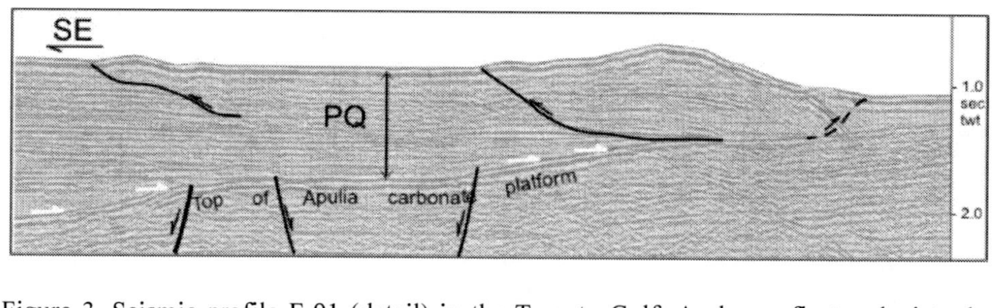

Figure 3. Seismic profile F-91 (detail) in the Taranto Gulf. A clear reflector depicts the base of the Plio-Quaternary sequence on-lapping (white arrows) the underlying Apulia carbonate platform, tilted towards the Calabrian Arc. The upper sediments are deformed by compressive tectonics, originating folding and back-thrusting, outside the external front of the chain, far-off about 10 kms.

Not far from these structures, along a seismic profile in a more southern position, we can observe (figure 4) another gentle back-thrust structure. It involves the shallow sedimentary sequence and clearly deforms the sea bottom, determining its uplift of about 100 meters. Some fault breakthroughs are highlighted by compressive reactivation of an extensive fault, cutting the base of the Plio-Quaternary sequence: this brings about evidence of positive inversion tectonics which represent relatively common structural elements on the studied area. These features are interpretable on different seismic reflection profiles, although their identification requires a very accurate observation and proper interpretation because of their unclear pinpointing. For instance it is possible to reconsider the above cited MS-26. A deeper analysis of data (figure 2b) reveals that the faults, although showing a predominant post-Miocene extension, are involved in compressive deformation of the shallow reflector which implies recent inversion tectonics. The vertical throws, produced by the former extensive deformation, remain the most evident tectonic feature interesting the post-Miocene sequence. In any case thrusting and mostly folding of the recent sediments clearly recall a compressive deformation.

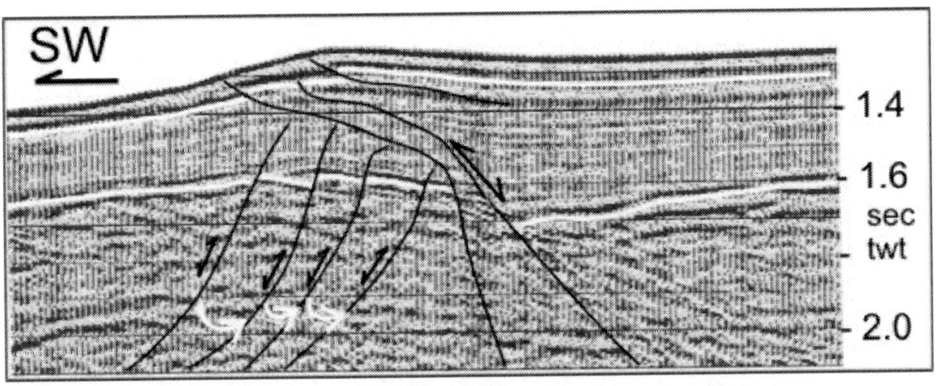

Figure 4. Seismic profile F-85 (detail) showing a positive tectonic inversion of a post-Messinian normal fault reactivated with a compressive style in Pleistocene time. The white deeper reflector represent the base of the Plio-Quaternary sequence, the upper white reflector, very recent, has been indicated to highlight the compressive structure producing some small thrusting. The sea bottom and the reflectors into the PQ sequence suggest that also the SW-verging faults are probably affected by the recent compressive stress with inversion tectonics, rotation of the intermediate blocks (highlighted by the white arrows) and a general uplift. Note that the positive structure is globally uplifted about 100 m relatively to the SW sea bottom.

As regards the time evolution of this inversion, the available data do not allow to make a detailed reconstruction. The Lieta-1 borehole, which is located about 26 km NW of the northern end of the profile (figure 1), below the sea bottom at 145 meters depth, crosses 365 meters of Pleistocene clays, and 39 meters of Lower Pliocene breccias and marls, which cover the Cretaceous Apulia platform. The calibration of some public profiles crossing the borehole and the MS-26, suggests the dating of the sin-extensive sediments in all likelihood Pliocene (Lower Pleistocene?) age while the sin-compressive sediments would be (Middle-Upper?) Pleistocene age.

A detailed analysis of some normal faults not reactivated by the inversion tectonics suggests an extensive activity gradually towards SE before the recent compressive phase. Neither here or in whole area, it is possible to identify the exact dating of the different deformational phases along the seismic profiles, because of the scarcity of boreholes and of their detailed analysis in the Pleistocene sequence.

As we can point out, the regional seismic dataset shows that a good part of the previous extensive faults, interesting the foreland during the post-Miocene time, are E-verging inward dipping faults, which are generally more inclined than the W-verging normal faults, and delimitate back-tilted fault blocks. Inversion tectonics are particularly evident just as reactivation of these faults, so to produce fault breakthrough with back-thrusts or, anyway, mostly W-verging fault-propagation folds at the tips of the previous normal faults.

In figure 5a, part of the CROP-M5 profile shows two apparently not inverted normal faults on the right, while a corrugated morphology can be seen of the left. In the detail shown in figure 5b, this morphology seems to be produced by some thrust inner the upper sediments, also involving the base of the Plio-Quaternary sequence, here interpreted as probable top of Messinian evaporite. The fault planes rapidly seem to flattening, probably it lies on the top of one of the Miocene, more consolidated, layers. The distance from the front of the Calabrian Arc, longer than 20 km (see figure 1), provides against the hypothesis that this thrust features can be attributable to the outer upper compression of the chain, but their origin have to be searched in deeper level, as later suggested.

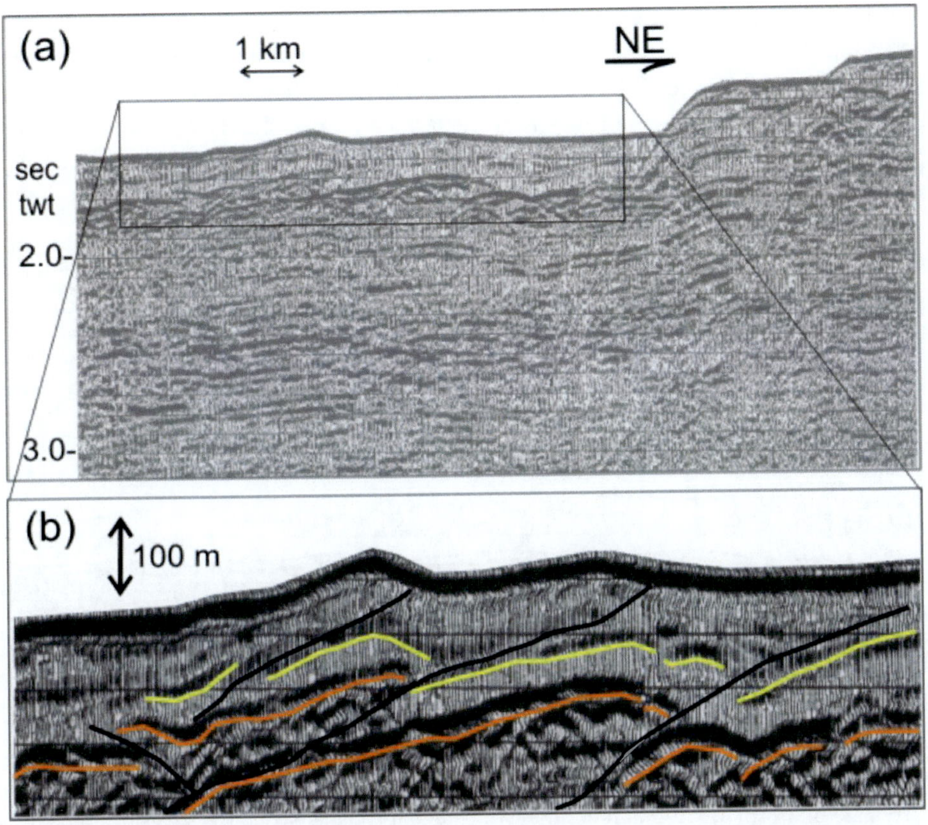

Figure 5. (a) Part of the seismic profile CROP-M5 (position in figure 1). The sea floor, cut by two normal faults on the right, appears also involved in some folds. These last, analyzed in better detail, result to be connected to compressive tectonics originating thrusting of the upper soft sediments and of the base (orange line) of Plio-Quaternary sequence. Note that these structures, very gentle with respect to the geological features generally interpreted in the seismic profiles, are long about 1 km and high some tens of meters (vertical scale for the sea bottom in meters is indicated).

Analyzing the seismic profiles across the seismic region, we can identify an area involved in a large number of faults with relatively small throw (figure 6a). Many of these faults converge on the deep Permo-Triassic (seismic detail of top and base of the sequence in figure 6b) and Upper Triassic-Jurassic layers, where the main seismic events are originated (Castello et al., 2006). In a first interpretation, these faults appear mainly extensive, and surely they cut the pre-Miocene sequence through a normal fault systems. Nevertheless, by observing the profiles of the area (figure 6c), we can highlight a characteristic recurrent morphology in this seismic area: mounded shallow sediments generally fill and cover the NW-SE oriented graben and the relative structural minima of the pre-Pliocene sequence. The shallow structures are generally clearly evident on the sea bottom profiles and on the parallel folded reflector inner the more recent sequence, probably deposited during main the Pleistocene age.

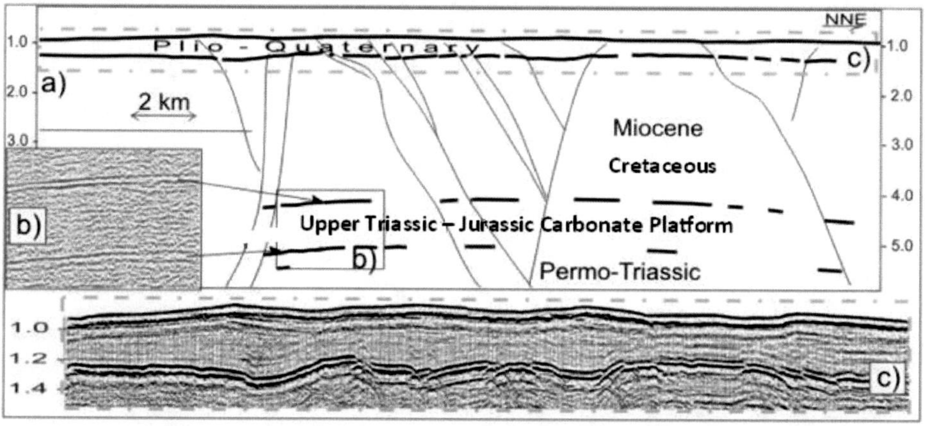

Figure 6. Part of the seismic profile F-55, crossing the seismic area of the South Salento offshore with a SSW-NNE direction (position in figure 1). a): Schematic line-drawing: only the sea bottom, the base of the Plio-Quaternary sequence, the top and the base of the carbonate platform are highlighted: the Miocene-Cretaceous sequence, even if affected by many multiples, seems anyway characterized by a stratified seismic facies, probably pertaining to a pelagic domain. This last covers the Upper Triassic-Jurassic carbonate platform, with evident top and bottom reflectors (b). The main faults affecting the base of the Plio-Quaternary sequence have been indicated, probably reactivating some older (Mesozoic) faults; the seismic events are positioned in the area crossed by this profile, with a depth ranging between 10 and 30 kms (Castello et al., 2006) corresponding to the thickness of Permo-Triassic sediments and of basement. In c) the upper part of the seismic profile has been vertically exaggerated (v.e. x6) to better show the typical swelling of the most recent horizons, interpreted as originated by compressive tectonics, focused above the post-Miocene grabens, bounded by normal faults.

This swelling is not very marked in the scale and resolution of seismic profiles; a vertical exaggeration can help analyze this kind of structures: in figure 7a we focused our attention on a typical graben described by the base of the Plio-Quaternary sequence which is cut by a system of quasi-symmetric normal faults. We can interpret (figure 7b) some sin-extension sub-horizontal reflectors (green) covered by two or three mounded reflectors (brown), including the sea bottom. These are interpretable as detachment fold related to a distributed deformation above the sin-extension/sin-compression boundary. Sometimes, a more marked inclination of the upper extension of the faults reminds of the inversion tectonics interpreted on the north-western area.

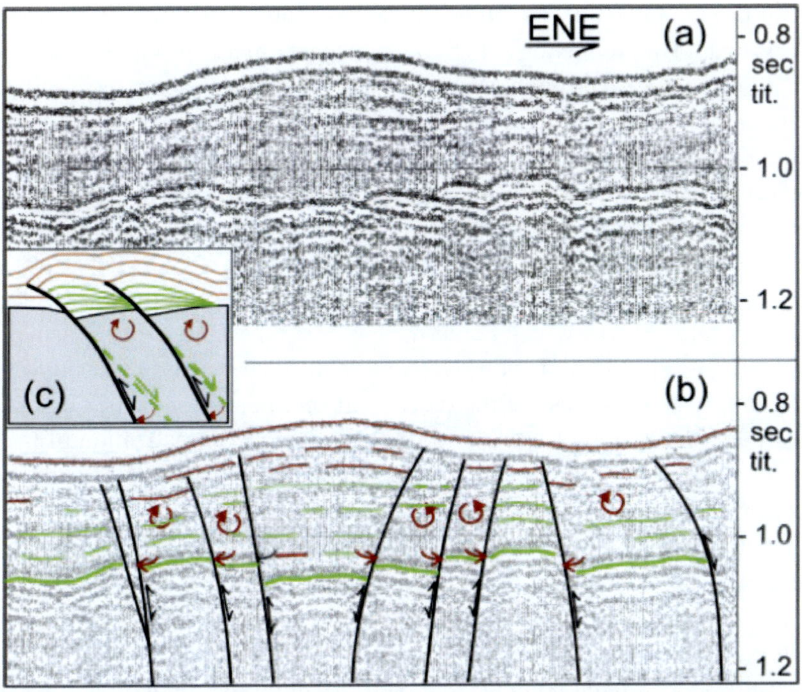

Figure 7. (a) Detail of the seismic profile D-475, position in figure 1. In (b) the base of the Plio-Quaternary sequence below 1 sec twt (green bold line) and other sin-extension reflectors (green thin lines) have been identified between the normal faults separating the partially tilted fault blocks. A more recent compressive deformation is identifiable by the folded upper reflectors (brown lines). that can be explained by (see also (c), after Wang et al., 1995) back rotation of the faults (brown arrows), verticalization and resultant uplift of the fault blocks, squeezing, folding and inverse faulting of the upper soft sediments due to the diminished length of the base.

Seismic stratigraphy and regional interpretation exclude the attribution of this shallow geometry to depositional or erosive features. Therefore an alternative hypothesis has to be proposed for this type of structuring. As well, we reconsidered the analogies with the above mentioned northern compressive structures: the whole data suggest that the swelling of the shallow sediments can be explained in the context of the compressive stress domain involving the Adria crust: this tectonic regime might have produced a rotation of the fault planes in a more vertical arrangement and the following back rotation of the previously tilted blocks (figure 7c). Following Wang et al. (1995), that interpreted an analogous deformation in the East China Sea Basin, this back rotation of the more rigid thickness during inversion tectonics, causes a reduced length of the upper soft sediment bed. The reduction of the disposable space produces compressive deformation, achieved through thrust sense reactivation along the fault planes. Sometimes the sin-rift fill are squeezed out into a series of folds and thrusts, as can be in figure 5; other times, still agreeing with Wang et al. (1995), the upper sin-compressive sequence detaches along a plane close to the sin-rift/post-rift boundary, as in figures 6 and 7, or they fold detaching along a plane separating soft by rigid lithologies, as in figure 2b.

In a more accurate analysis of figure 4, we can interpret the back-rotation of the previous tilted blocks which could have produced a more vertical arrangement of the deeper rigid blocks. This movement should have caused their uplift and a diminished length of the base of the recent soft sediments, forced to a compressive deformation evidenced by folding and back-thrusting. Verticalization of the lower rigid blocks and deformation of the soft upper sequence has totally produced over 100 meters of uplift.

It must be point out that these swelling structures, generally very gentle, can be interpreted also to the East of the seismic area, where the Adria foreland began deepening towards the front of the Hellenic Arc, far off of only about 40 kilometers.

Therefore, the main compressive deformation, interpreted on the seismic profiles, seems correlated to the seismicity of the South Salento offshore and particularly to the few available focal mechanisms of the area. The extensional tectonics, associated to the hinge of the Adria bulge and/or to the inner arc invoked by many authors (Doglioni et al., 1994; Argnani et al., 2001; Billi and Salvini, 2003), seem to be attributable to a previous phase, nowadays replaced by compressive tectonics. The so produced back rotation of the normal faults would have involved the whole brittle crustal thickness of the foreland. By this way, the fault blocks verticalization probably would play an important part on the bulge, contributing to the uplift of the Adria foreland.

The focal mechanisms also suggest a strike-slip components that, hardly recognizable along the seismic profiles, should involve the NW-SE fault system with a main left horizontal throw, and therefore contributing to their reactivation. By the observation of the seismic profiles, we can only hypothesize an horizontal displacement along the active faults, which probably converge on few sub-vertical crustal faults (see, for instance, figure 6) in a kind of flower structures. The strike-slip tectonics strongly involved the Calabrian Arc, due to the oblique convergence between the Calabrian/South Apennine Chain and the western Apulia margin. Del Ben et al. (2008a), on the base of time evolution of the strike-slip systems of the arc, found out a changing direction of its migration, from ESE-ward until Lower Pleistocene age, to SSE-ward since the Middle Pleistocene age, with a left strike-slip system along the Apulia margin. The collision of the chain with the thicker continental Adria foreland in the west-Salento offshore, probably transferred this left strike-slip component also to the NW-SE fault system of the foreland; this is likely to be occurred especially since the Middle Pleistocene time, when the Calabrian Arc assumed a SSE-ward migration, parallel to the pre-existing normal faults of the Adria.

In this context, we can suppose that the inversion tectonics, combined to the strike-slip component, were gradually SE propagating during the Plio-Quaternary age, reaching the Salento offshore in recent time.

Conclusion

The western Apulia margin represents a resistant foreland of the Calabrian Arc migrating toward the still existing oceanic Ionian foreland. The eastern margin is instead involved in the compressive system of the Hellenic Chain. This double and nearly opposite chains, migrating toward and above the Adria foreland, determines its bending with probable partial re-activation of the Mesozoic normal faults, parallel to the Apulia margins.

The extensive activity along the western margin developed mostly during the Upper Pliocene/Lower Pleistocene age and, only gradually in more recent times, toward the SE. They were likely to be activated during the subduction of the oceanic Ionian crust, which also pulled down the adjoining continental Adria plate. The Adria tilting originated the normal faults, often inward dipping and producing important back-tilting of the fault blocks, resulting in strongly inclined fault planes, generally more pronounced than the W-verging faults.

The oblique collision of the Calabrian Arc with the continental Adria plate, probably during Early Pleistocene age, determined a slower migration of the chain and the decrease of extensive tectonics in the Marsili Basin.

Since the Middle Pleistocene age, along the western margin of the continental foreland, the Calabrian Arc changed its migration toward SSE, sliding with a left transpressive tectonics along the same margin. The nearly opposite compression from the Hellenic Chain transmitted the stress regime to the foreland and induced a back rotation of the previously tilted fault blocks, so to turn into a more vertical arrangement which produced an uplifting of the rigid thickness. Furthermore, the base of the upper soft sediments was subjected to change length, prompting detachment folds, or fault-propagation folds, or fault breakthrough by compressive reactivation of the previous extensional faults. These compressive features, although very gentle in comparison to the seismic resolution of the interpreted profiles, show an uplifted profile which generally reaches some tens of meters, sometimes exceeding 100 meters. The sea bottom never on-laps these structures, so suggesting that this deformation has to be assigned to the present tectonic activity.

These positive inversion tectonics seem to be favourished by more inclined normal faults, often connected to the back-tilting processes. In this way, along the west-Apulia margin, the recent compressive deformation is mainly related to back-thrust or back-fold structures.

The depth of the seismic events suggests that the compressive regime would not be induced by the outer front of the Calabrian and Hellenic Chains, but by the deep collision of the chains with the continental Adria foreland, resistant to the subduction. This stress regime reactivated previous faults in a weaker crustal zone, as suggested by transition from platform to pelagic sequences in the area of seismic events. The back-rotation of the fault blocks would have involved the crustal thickness of the Adria plate, contributing to the regional uplift generally attributed to a forebulge.

The oblique convergence between the Calabrian/South Apennine Chain and the Adria margin might have determined a left strike-slip component transmitted to the foreland, along the NW-SE fault system, as pointed out by the focal mechanisms. This component, SE propagating during the Plio-Quaternary age, simultaneously with the Calabrian Arc migration, reached the study area in recent time, probably emphasized in the Middle Pleistocene time, when the direction of the arc started moving SSE-ward.

Finally, the available data show that the recent and relatively low compressive stress on the South Salento offshore originated some gentle structures, which are visible only by an accurate observation of the seismic profiles, generally describing features of more regional scale. This hardness suggests that perhaps also in the onshore of the Salento peninsula could exist some structures due to inversion tectonics, but back rotation of the rigid blocks is not easy to recognize, and soft sediments are probably too scarce to reveal thrust and folds. Therefore, inversion tectonics could be not still highlighted.

References

Argnani, A., Frugoni, F., Cosi, R., Ligi, M. & Favali, P. (2001). Tectonics and seismicity of the Apulian Ridge south of Salento peninsula (Southern Italy) *Annali di Geofisica.* Vol. 44, pp. 527-540.

Armigliato, A., Tinti, S., Pagnoni, G., Mastronuzzi, G., Pignatelli, C., Sansò, P., Gallazzi, S., Tonini, R. & Brizuela Reyes, B. (2007). Studio preliminare sulla sorgente del terremoto del 20 Febbraio 1743 in Puglia Meridionale a partire da dati di tsunami. Convegno Nazionale di Geofisica della Terra Solida. Trieste, 6-8 Ottobre 2008. Riassunti Estesi (pp. 36-37). ISBN/ISSN: 88-902101-3-3.

Aubouin, J., Blanchet, R., Cadet, J.P., Chavret, J., Chorowics, J., Cousin, M., & Rampnoux, J.P. (1970). Essai sur la geologie de Dinarides. *Bull. Soc. Geol. Fr.* Vol. 12 (7), pp.1060-1095.

Billi, A., & Salvini, F. (2003). Development of systematic joints in response to flexure-related fibre stress in flexed foreland plates: the Apulian forebulge case history, Italy. *Journal of Geodynamics.* Vol. 36, pp. 523-536.

Bosellini, F., Russo, A., & Vescogni, A. (2001). Messinian reef-building assemblages of the Salento Peninsula (southern Italy): palaeobathymetric and palaeoclimatic significance. *Palaeogeography, Palaeoclimatology, Palaeoecology*. Vol. 175, pp. 7-26.

Castello, B., Selvaggi, G., Chiarabba, C., Amato, A. (2006). CSI Catalogo della sismicità italiana 1981-2002, versione 1.1. INGV-CNT, Roma Bollettino Sismico Italiano, Istituto Nazionale di Geofisica e Vulcanologia Doumaz F. & Vinci S. Portale di divulgazione e diffusione di dati cartografici digitali del Centro Nazionale Terremoti (INGV) - http://kharita.rm.ingv.it/dmap/

de Alteriis, G., & Aiello, G. (1993). Stratigraphy and tectonics offshore of Puglia (Italy, southern Adriatic Sea) *Marine Geology.* Vol. 113, pp. 233-253.

Del Ben, A., Barnaba, C., & Taboga, A. (2008a). Strike-slip systems as the main tectonic features in the Plio-Quaternary kinematics of the Calabrian Arc. Marine Geophysical Research Vol. 29, pp. 1-12, DOI 10.1007/s11001-007-9041-6.

Del Ben, A., Geletti, R., & Mocnik, A. (2008b). Relation between recent tectonics and Mesozoic inherited structures of the central-southern Adria Plate. Convegno Nazionale GNGTS. Trieste, 6-8 Ottobre 2008. Riassunti Estesi (pp. 25-26). ISBN/ISSN: 88-902101-3-3.

D'Ingeo, F., Calcagnile, G., & Panza, G.F. (1980). On the fault plane solutions in the Central-Eastern Mediterranean region. *Boll. Geofis. Teor. Appl.*, **22**, 13-22.Doglioni C., Mongelli F. and Pieri P. (1994). The Puglia uplift (SE Italy):

an anomaly in the foreland of the Apenninic subduction due to buckling of a thick continental lithosphere. *Tectonics.* Vol. 13, pp. 1309-1321.

Doglioni C., Mongelli F., & Pieri P. (1994) The Puglia uplift (SE Italy): An anomaly in the foreland of the Apenninic subduction due to buckling of a thick continental lithosphere. Tectonics. Vol. 13, pp. 1309-1321.

Favali P., Mele G., & Mattietti G. (1990). Contribution to the study of the Apulian Microplate Geodinamics. *Mem. Soc. Geol. It.* Vol. 44, pp. 71-80.

Finetti, I.R.; Ed.; (2005). CROP Project-Deep Seismic Exploration of the Central Mediterranean and Italy, Vol. 1, ISBN: 0-444-50693-4. Amsterdam: Elsevier (Netherlands).

Gambini, R., & Tozzi, M. (1996). Tertiary geodynamic evolution of the Southern Adria microplate. *Terra Nova.* Vol. 8, pp. 593-602.

Mastronuzzi, G., Pignatelli, C., Sansò, P., & Selleri, G. (2006). Boulder accumulations produced by the 20th of February, 1743 tsunami along the coast of southeastern Salento (Apulia region, Italy). *Mar. Geol.* doi:10.1016/j.margeo.2006.10.025

Mattavelli, & Novelli (1990). Geochemistry habitat of the oils in Italy. *AAPG.* Vol. 74, pp. 1623-1639.

Mattavelli, L., Novelli, L., & Anelli, L. (1991). Occurrence of hydrocarbons in the Adriatic basin. *Generation, accumulation and production of Europe's hydrocarbons, Special Pubblication of the European Association of Petroleum Geoscientists,* Spencer A.M (ed.), pp. 369-380.

Nicolai, C., & Gambini, R. (2007). Structural architecture of the Adria platform-and-basin system. *Boll.Soc.Geol.It.* Vol. 7, pp. 21-37.

Nicolosi, I., Speranza, F., & Chiappini, M. (2006). Ultrafast oceanic spreading of the Marsili Basin, southern Tyrrhenian Sea: Evidence from magnetic anomaly analysis. *Geology.* Vol. 34, pp. 712-720.

Patacca, E., Scandone, P., Di Luzio, E., & Cavinato, G. (2008). Structural architecture of the central Apennines: Interpretation of the CROP 11 seismic profile from the Adriatic coast to the orographic divide. *Tectonics.* Vol. 27, TC3006, doi:10.1029/2005TC001917.

Royden, L. E. (1988). Flexural behaviour of the continental lithosphere in Italy: constraints imposed by gravity and deflection data. *J. Geophys. Res.* Vol. 93(B7), pp. 7747-7766.

Royden, L. E., Patacca, E., & Scandone, P. (1987). Segmentation and configuration of subducted lithosphere in Italy: an important control on thrust-belt and foredeep basin evolution. *Geology.* Vol. 15, pp. 714-717.

Scrocca, D., Doglioni, C., Innocenti, F., Manetti, P., Mazzotti, A., Bertelli, L., Burbi, L., & D'Offizi, S.; Eds; (2003). CROP atlas -seismic reflection profiles of the Italian crust. *Memorie descrittive della Carta Geologica d'Italia.* Vol. 52.

Shaw, J.H., Connors, C., & Suppe, J.; Eds; (2005). Seismic Interpretation of Contractional Fault-Related Folds. *AAPG Seismic Atlas, Studies in Geology.* Vol. 53, Tulsa, Oklahoma, USA.

Tinti, S., Maramai, A., & Graziani, L. (2004). *The New Catalog of Italian Tsunamis. Nat. Haz.,* Vol. 33, pp. 439-465.

Tozzi, M., Kissel, C., Funicello, R., Laj, C., & Parotto, M. (1988). A clockwise rotation of Southern Apulia? *Geophys. Res. Lett.* Vol. 15, pp. 681-684.

Vannucci, G., Pondrelli, S., Argnani, A., Morelli, A., Gasperini, P., & Boschi, E. (2004). *An Atlas of Mediterranean Seismicity Annals of Geophysics,* suppl. to Vol. 47.

Wang, G.M., Coward, M.P., Yuan, W., Liu, S., & Wang, W. (1995). Fold growth during basin inversion – example from the East China Sea Basin. Buchanan J. G. and Buchanan P.G. (eds) *Basin Inversion, Geology Society Sp. Publ.* No.88, pp. 493-522.

In: Recent Progress on Earthquake Geology ISBN: 978-1-60876-147-0
Editor: Pierpaolo Guarnieri, pp. 165-187

Chapter 7

THE CONTRIBUTION OF STRUCTURAL GEOLOGY AND REGIONAL TECTONICS TO THE DEFINITION OF LARGE-SCALE SEISMOTECTONIC PROVINCES AND INDIVIDUAL SEISMOGENIC SOURCES: APPLICATION TO THE EXTENSIONAL BELT OF CENTRAL ITALY

***Giusy Lavecchia*[1]*, Paolo Boncio*[1]*, Francesco Brozzetti*[1]*, Rita de Nardis*[2] *and Francesco Visini*[1]**

[1]Laboratorio di Geodinamica e Sismogenesi (GEOSIS-LAB) - Dipartimento di Scienze della Terra, Ud'A, Campus Universitario, 66013, Chieti, Italy
[2] Dipartimento della Protezione Civile, Rome, Italy

ABSTRACT

The essential contribution of structural geology to the identification and parameterization of active faults responsible for moderate to large earthquakes is now largely recognised. When both geological and good-quality instrumental seismological data are available, the geometry, kinematics and slip rate of the individual seismogenic source can be rather well constrained. The definition of the 3D shape and size of an active fault segment, which is fundamental for seismic hazard assessment purposes, is evidently more difficult when only historically activated and/or silent sources are involved. In such a case, we can benefit from the combined use of structural geology and regional tectonics to determine a long-term geologic and kinematic context for the present deformation field and associated seismicity.

The integration of surface and deep structural data with historical and instrumental seismicity has allowed us to define the 3D geometry of the intra-Apennine extensional seismotectonic province in central Italy. The province coincides with a regional seismogenic fault system, which extends for nearly 500 km from Lunigiana, in northern Tuscany, to Sangro Valley in southern Abruzzo. It consists of WSW- to SSW-dipping high-angle normal and normal-oblique faults developed at upper- to mid-crust depths within the hangingwall of an E-dipping low-angle detachment system, named the Etrurian Fault System. We have defined a segmentation model consisting of major individual fault segments separated by first-order structural complexities and discontinuities, which would act as a potential barrier to the propagation of major earthquake ruptures. We have named the plan projection of such major active structures "seismogenic boxes" and evaluated the maximum expected earthquake for each of them. Within the area considered hereunder, we have identified and parameterised 37 boxes that are responsible for or capable of experiencing normal-faulting earthquakes with magnitude ≥ 5.5. Most of these sources are associated with moderate-angle (on average 50°) west-dipping faults, but a few of them are interestingly associated with low-angle (25°–30°) east-dipping sources. The overall integrated analysis is aimed at improving the understanding of both the structural style and the seismogenic potential of the active extensional deformation in Italy for seismic hazard purposes.

Keywords: central Italy, active extension, seismicity, seismogenic source, seismotectonic zonation

INTRODUCTION

Advanced knowledge of the active deformation field and seismotectonics is to be recognised in Italy, thanks to a number of multidisciplinary projects at national level, as well as detailed works on individual fault sources (Meletti et al., 2000 and 2008 reference therein). The kinematic approach has been commonly used in identifying homogenous seismotectonic zones. This means that the zones have been defined on the basis of 3D structural-geological and earthquake distribution data, also integrated with regional data on to the Pliocene-Quaternary history of geological deformation. Together with the definition of the regional scale zones, a database of individual seismogenic sources responsible for major earthquakes, known as DISS, has been developed and progressively updated by scientists of the INGV (Istituto Nazionale di Geofisica e Vulcanologia) (Valensise and Pantosti, 2001; Basili et al., 2008). The DISS database combined seismogenic sources purely defined on macroseismic data, with seismogenic sources defined on a geological-geophysical basis.

A detailed structural-seismotectonic approach to the zonation of central Italy has also been developed by the GEOSIS-LAB research group in Chieti (e.g., Lavecchia et al. 1994, 2002; Boncio et al., 2004; Pace et al., 2006). The surface geology data have been integrated with seismological and subsurface structural data to define the 3D geometry of the active fault structures and their seismogenic role. The expression "seismogenic boxes" has been used to refer to the surface projection of individual active faults associated with the largest earthquakes of a region. Simultaneously, 3D structural-kinematic criteria have been used to delimit wider domains, which are kinematically and seismotectonically homogenous - e.g. the seismotectonic provinces - and also include the minor seismicity lying outside the boxes. Lavecchia et al. (2002) and Basili et al. (2008) also identified another type of seismogenic source, here called "seismogenic alignment", which may be located in an intermediate conceptual position between the seismotectonic province and the seismogenic box; in fact, it is the surface representation of a fault system that spans an unspecific number of individual sources.

The use of the individual seismogenic structures, defined by means of structural-geological criteria, might be helpful in reconstructing the long-term seismic cycle and, consequently, in estimating the seismic hazard of areas where the earthquake catalogues could be insufficient. Moreover, it could also favour the transition from poissonian, stationary in time, to time-dependent probabilistic seismic hazard assessments (PSHA), which are sensitive at the time elapsed since the last meaningful earthquake (Pace et al., 2006).

Since it is not always possible to identify and parameterise the individual seismogenic structures on a geological basis, the available multidisciplinary information leads to the development of a "multi-layered" zonation (Pace et al., 2006), i.e. a number of geologically-constrained source-layers responsible for the seismicity of a region above a given magnitude threshold. For example, the following classification is possible: (a) a basal layer containing large polygons corresponding to regional-scale "seismotectonic provinces", (b) an intermediate layer with long and narrow seismic areas corresponding to the "seismogenic alignments", and (c) an upper layer consisting of rectangular-shaped seismogenic boxes. We might also consider another layer, not shaped on a geological-structural basis, containing the background instrumental seismicity.

In this chapter, a brief definition of the different sources that can be used to construct the multilayer zonation of the extensional belt of central Italy will be provided. Then, the geometry of the late Pliocene-Quaternary extensional structures and the related seismicity will be introduced, with special emphasis on the low-angle east-dipping seismogenic system. Finally, the multilayered zonation will be presented, describing the different types of identified sources and their association with major earthquakes. Lastly, results and considerations will be summarised.

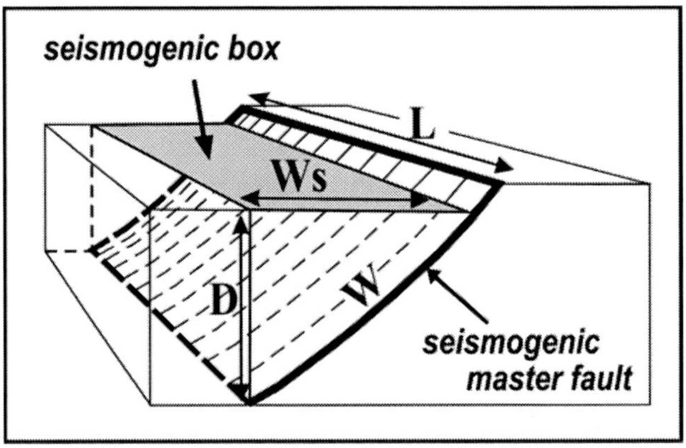

Figure 1. Three-dimensional sketch of a seismogenic box (from Lavecchia et al., 2002): L = along-strike length of the seismogenic master fault; W = down-dip width of the seismogenic master fault; Ws = surface box width; D = thickness of the local seismogenic layer (i.e. maximum depth-extent of the seismogenic fault plane).

Types of Seismotectonic Domains and Seismogenic Sources

Seismotectonic Province

The expression "seismotectonic province" (SP) refers to a large structural domain which is homogeneous from the point of view of (1) the Late Pliocene-Quaternary kinematics and stress field, (2) the crustal structure (Moho thickness, Bouguer anomaly, heat flow), and (3) the type and level of the seismic activity. The province's surface boundaries do not simply include an area with the above-mentioned characteristics, but are also drawn to give a 2D-view of a 3D regional-scale seismogenic volume. Therefore they coincide either with outcropping linear tectonic elements (e.g. an active thrust front or an extensional breakaway fault zone) or with the surface projection of the intersection line between two deep planar elements (e.g. the intersection between a seismogenic tectonic discontinuity with the brittle ductile-transition or with another- inactive- tectonic element which can act as a barrier). Any known earthquake, above an assumed magnitude threshold ($M_w \geq 4.5$ in the present case), which lies within the boundary of the province, must be given a consistent tectonic collocation and interpretation.

Seismogenic Alignment

The expression "seismogenic alignment" (SA) refers to a seismic area that is some kilometres wide and some tens of kilometres long and is the surface projection of the seismogenic rock volume at the hangingwall of an active fault alignment. The latter is the along-strike envelope of neighbouring individual seismogenic master faults, associated with historical and/or instrumental earthquakes above a given threshold of magnitude ($M_w \geq 5.5$ in the present case). The SA surface width, which is the projection of the fault down-dip width, may vary along strike depending on the change in dip of the individual faults and the change in thickness of the seismogenic layer. Within each SA, this variation would be moderate, because it is one of the three factors controlling the maximum expected earthquake.

Seismogenic Master Fault and Seismogenic Box

The expression "seismogenic master fault" (SMF) refers to an individual major Quaternary fault, continuous along strike for several kilometres and responsible for earthquakes of relevant magnitude ($M_w \geq 5.5$ in the present case). Each SMF has to meet one or more of the following requirements: (1) evidence of activation during instrumental seismic sequences and/or clear association with historical earthquakes; (2) evidence of paleo-seismological activity; (3) evidence of activity during the Late Pleistocene-Holocene time; (4) presence of along-strike neighbouring faults, which meet one or more of the above requirements.

In order to define individual SMF segments within a regional alignment of the active structures, we consider segment boundaries such as fault gaps of more than 3–4 km between aligned structures, *en échelon* step-overs with steps of more than 3 to 5 km and abrupt along strike fault bends. These types of first order structural complexities are assumed to act as barriers, preventing or stopping the rupture propagation during an earthquake. Complexities of minor hierarchical order (e.g. hundreds of metres step-overs), which in some cases interrupt the SMF fault continuity on the surface, are not considered earthquake segment boundaries, as they would be not able to interrupt the fault continuity at depth.

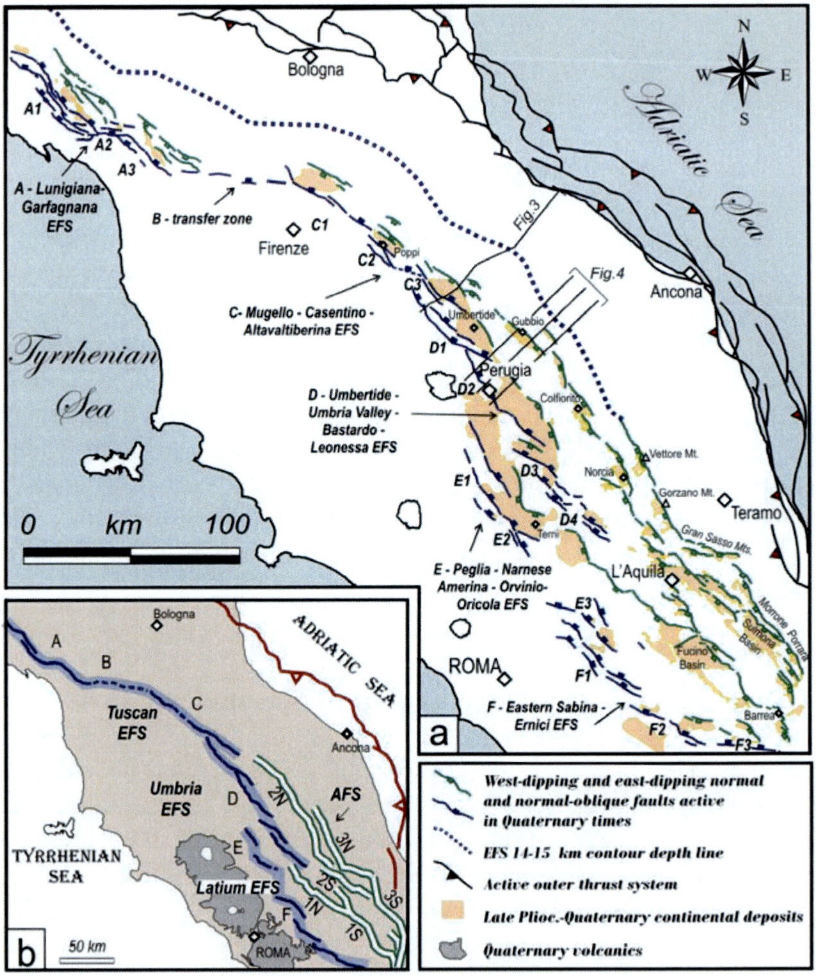

Figure 2. (a) Structural map of the Late Pliocene-Quaternary extensional belt in northern Italy with traces of the geological sections of Figures 3 and 4 (modified from Lavecchia et al., 2002; Boncio et al., 2000 and 2004; Brozzetti et al., 2009). (b) East-dipping (in blue) and west-dipping (in green) seismogenic fault alignments belonging to the Etrurian Fault System (EFS) and to the Apennine Fault System (AFS), respectively.

For the maximum expected magnitude of each SMF to be inferred, the area of the fault plane must be known. Schematically, this can be assumed to be equal to the along-strike length (L) by the down-dip width (W). L can be constrained by direct geological and/or morphotectonic field-mapping, while W can be derived from indirect active and passive seismic data (e.g. seismic reflection profiles and depth distribution of seismicity), as well as from rheologic considerations.

Once the shape (listric or planar), the attitude (strike and dip) and the size (L, W) of any given SMF have been defined, its 3D geometry can be schematically represented in a map view as a small rectangular-shaped area, named "seismogenic box" (Figure 1). The box length (L) coincides with the surface fault length; the box width coincides with the surface projection (W_s) of the fault width. W_s depends on both the dip angle of the seismogenic fault plane and the depth (D) of the local seismogenic layer. The dimension of the box in map view is proportionate to the area of the seismogenic fault and, therefore, to the seismic moment predictable for the major earthquake on the fault. The definition of the average slip rate for each SMF and its association with a defined number of known earthquakes give additional basic information to estimate the seismicity rates and the recurrence time of the maximum expected earthquake.

The Active Extensional Belt of Central Italy

The Apennines in central Italy are a NE-verging fold-and-thrust system developed in Mio-Pliocene times, in the course of an eastward migrating contractional process. Since the Late Miocene, shortening has been accomplished by extension in the back, with nucleation of normal fault systems migrating eastward, as well (Lavecchia et al., 1994). The outermost extensional belt started to develop in the Late Pliocene (ca. 3.5 Ma) and is still active. In central Italy, it extends for nearly 500 km in an average NW-SE direction along the axis of the present mountain chain (Figure 2a). It consists of both east-dipping, prevailingly low-angle, and west-dipping, prevailingly high-angle, normal fault systems.

In this chapter, the general names of Etrurian Fault System (EFS) and of Apennine Fault System (AFS) will be given the former and the latter, respectively. The mutual relations between the structures of the two systems, and the prevalence of one of the two in terms of control on the basins growth and/or on the present seismic activity, change along the strike of the belt. The integrated analysis of geological and geophysical data show that the east-dipping system prevails in the Tuscan and western Umbria areas. It controls the Pleistocene-Holocene basins evolution, accommodating most of the extensional deformations, and represents the basal detachment of the west-dipping structures (Barchi et al., 1999; Boncio et al., 2000). In the eastern Umbria-Marche-Abruzzo area, the west-dipping faults clearly prevail, both numerically and in terms of accommodated offset and associated earthquakes (Galadini and Galli, 2000; Boncio et al., 2004 with references). The west-dipping faults are well exposed and well known in the geological cartography and literature. Typical and representative faults belonging to the AFS are the Norcia and Colfiorito faults in the Umbria-Marche region and the Fucino fault in the Abruzzo region. They were all activated by destructive events during the 20th century (see seismogenic source database at http://diss.rm.ingv.it/diss/ with references).

The east-dipping system has only recently been identified thanks to field mapping and geological section balancing, integrated with the interpretation of seismic lines, also constrained through deep-well stratigraphy. The EFS has been accurately investigated especially across the Val Tiberina and Umbria Valley of the Umbria region (Brozzetti, 1995; Barchi et al., 1998; Boncio et al., 2000; Collettini and Barchi, 2002; Brozzetti et al., 2009), but seismic lines are available and partially interpreted also in Tuscany, across the Lunigiana-Garfagnana area (Argnani et al., 2003; Brozzetti et al., 2007). The interpretative sections across the EFS show that the entire system cross-cuts and offsets the previous fold-and-thrust structures (Figure 3). They also show a major low-angle normal fault, which slopes beneath the Apennine chain to depths of 14-15 km and is the basal detachment for synthetic east-dipping and antithetic west-dipping faults, which splay at different depth values from it. The eastward-dipping low-angle contact often downthrows the Ligurian Units, previously overthrusted on the Tuscan Units, and/or displaces late Miocene terrains above Triassic evaporites with the elision of the Meso-Cenozoic sequence. Asymmetric grabens with a steeper south-western flank, filled by a sequence of fluvio-lacustrine deposits of Pliocene - Pleistocene age, are developed at the EFS hangingwall, in the vicinity of the surface breakaway fault zone. These basins (Lunigiana, Garfagnana, Mugello, Casentino and Val Tiberina), located at the hangingwall of the EFS, are all similar in surface geology and morphology, and are clearly controlled by the NE–dipping master faults, whereas the antithetic SW-dipping faults are second-order structures (Benvenuti, 1997; Boncio et al., 2000; Bernini and Papani, 2002; Argnani et al., 2003; Benvenuti, 2003; Brozzetti et al., 2009).

From north to south, the EFS of central Italy can be divided into three major right stepping branches (Figure 2b). The northern branch is mainly located within the Tuscany territory (Tuscan EFS): it extends with an average N140°E strike for nearly 240 km and consists of two major fault systems and the related basins (Lunigiana-Garfagnana, A, and Mugello-Casentino-Valtiberina, C), separated by a badly exposed E-W transfer zone along the Bagni di Lucca – Barberino alignment (B). The central branch, mainly located in the Umbria territory, (Umbria EFS) extends along strike for nearly 160 km and consists of the four major fault systems, which bound to the west the Umbertide, Umbria Valley, Bastardo, Monteleone and Leonessa basins (D_1-D_4). The east-dipping faults of the Val Tiberina basin (Città di Castello and Anghiari faults) are the southward end of the Tuscan EFS, whereas the faults located nearly 10–15 km westward of the basin border (Mt. Favalto and M. Santa Maria Tiberina faults) are the northward end of the Umbria EFS. The southern EFS branch is mainly located in the Latium territory (Latium EFS). It is currently less defined than the others and consists of two major fault systems, one bounding the M. Peglia-Terni basins (E_1, E_2) to the east and the other developing across the eastern Sabina-M. Ernici area (F_1, F_2).

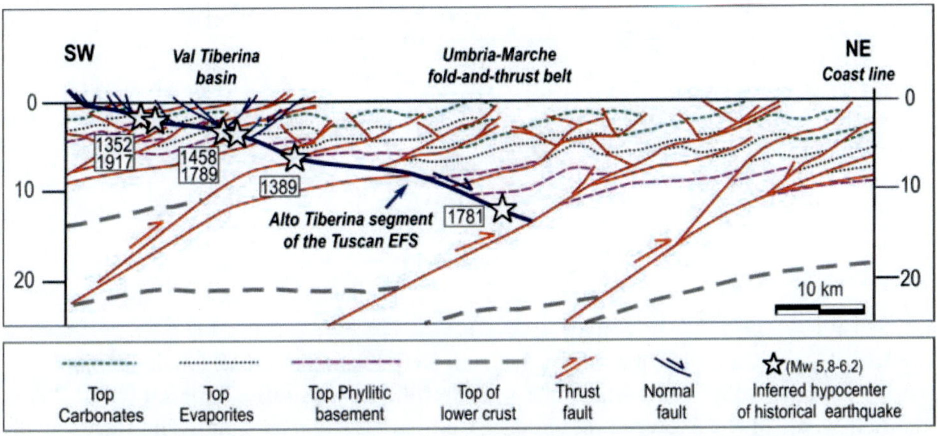

Figure 3. Geological interpretation of a portion of the CROP 03 near the vertical seismic line (after Lavecchia et al., 2003). The trace of the section is given in Figure 2a. The stars above the EFS basal detachment plane indicate historical events possibly associated with it, as hypothesised by Brozzetti et al. (2009).

Available focal mechanisms (CMT, Regional CMT and P wave first-motion solutions; Pondrelli et al., 2006) indicate a prevalence, all over the central Italy extensional belt, of normal and normal-oblique kinematics, with T axes oriented SW–NE to WSW–ENE (Figure 5b). The kinematics indicates an average SW–NE stretching direction, which is consistent with the pattern of recent and active extensional faulting obtained from geological data, and is also consistent with geodetic data, which indicate active NE-trending stretching at rates of 2.5 mm/yr

(Hunstad et al., 2003; D'Agostino et al., 2008), as well as with active stress determinations (Boncio and Lavecchia, 2000; Montone et al., 2004).

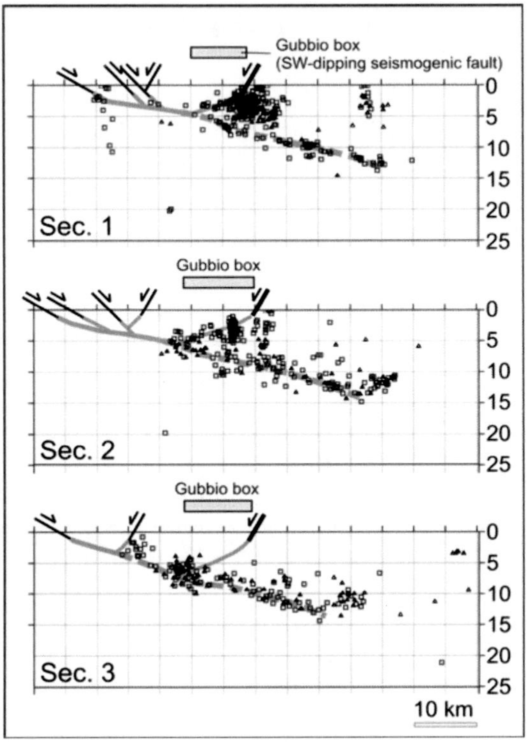

Figure 4. Hypocentral distributions of the accurately located microseismicity of the May–June 1987 and October 2000–May 2001 local seismic surveys (after Brozzetti et al., 2009). The half-width of the projected seismic belt on each section is 4 km. The traces of the sections are given in Figure 2a. The depth geometry of the EFS basal detachment is given for comparison.

The focal mechanisms and the aftershock sequences of the major instrumental earthquakes of the last 30 years (Norcia 1979 M_W 5.9, Gubbio 1984 M_W 5.6, Sangro Valley 1984 M_W 5.9, Colfiorito 1997 M_W 6.0), all associated with the west-dipping faults, show moderately-dipping planes (40–50°) joining at the surface steeply-dipping (50–70°) outcropping normal faults and locally reactivating at depth both low-angle (20–30°) and high-angle (70–75°) pre-existing faults (Boncio and Lavecchia, 2000; Pace et al., 2002; Chiaraluce et al., 2004, 2005; Pondrelli et al., 2006).

Rare focal mechanisms of minor events associated to the east-dipping faults show low-angle (20–30°) nodal plane dipping to the east (Brozzetti et al., 2009). Although most of the central Apennine seismicity occurs along the west-dipping faults, the available well-constrained microseismicity shows a wedge-shaped seismogenic volume, which progressively deepens eastward and is downward delimited by the EFS low-angle basal detachment (Boncio et al., 1998, 2000; Chiaraluce et al., 2007; Brozzetti et al., 2009) (Figure 4). Such a control of the EFS on the shape of the seismogenic volume is well evident across the Tuscan and Umbria branches and the northernmost fault system of the Latium branch, whereas it disappears southward.

In Figure 4, the hypocentres of well-constrained background seismicity are projected along three cross-sections, perpendicular to the Umbria EFS ($0.1 \leq M \leq 4.1$; May-June 1987 and Oct. 2000-May 2001 experiments; Boncio et al., 1998; Piccinini et al., 2003; Chiaraluce et al., 2007). The hypocentres span a depth range of more than 15 km and show a generalised eastward deepening consistent with the independently reconstructed geometry of the EFS basal detachment.

Seismogenic Alignments and Associated Earthquakes

The individual seismogenic sources identified within the central Italy extensional belt and responsible for $M_w \geq 5.5$ earthquakes are mainly associated to the listric and/or planar west-dipping faults of the Apennine Fault System (AFS) and, subordinately, to some of the low-angle (25–30°) east-dipping faults of the Etrurian Fault System (EFS). Fault-slip data, from the field mapping and the focal mechanism analysis, can be used to properly constrain the configuration of the active state of strain of both the AFS and EFS and to predict that the expected kinematics of the earthquakes is normal or normal-oblique with an average SW-NE directed tensional axis. Almost all the faults of the AFS identified in Figure 2a are well exposed and recognisable in the field. The maximum geological offset ranges from a few hundred meters to 2 km and the estimated minimum late Quaternary slip rates range from 0.3 to 1 mm/a (Galadini and Galli, 2000; Pace et al., 2006). Also the EFS shows evidence of recent-to-present activity supported by the imprinting on the morphologic and sedimentary evolution of the hangingwall block during Quaternary times, as well as by seismic reflection data coupled with instrumental seismicity (Boncio et al., 1998, 2000; Brozzetti et al., 2009). The evidence of the seismic activity and the seismogenic potential of these structures decrease southward, being well evident and consistent for the Tuscan EFS, subordinate for the Umbrian segment and hypothetic for the Latium branch. Interestingly, the Quaternary offset, as well as the evidence of seismic activity, decreases southward for the EFS and northward for the AFS.

Geological observations have allowed us to define a fault hierarchy within both the AFS and the EFS. Within the AFS, we have identified three regional fault alignments, hereunder named internal, intermediate and external (1, 2 and 3 in Figure 2b) and located in the Umbria-Marche-Abruzzo Apennine mountain chain. Within the EFS, five major right-stepping fault systems located in Tuscany, western Umbria and Latium can be identified. From north to south, they are the Lunigiana - North Apuane - Garfagnana EFS (A in Figure 2b), the Mugello - Casentino - San Sepolcro EFS (C in Figure 2b), the Umbertide-Umbria Valley-Leonessa EFS (D in Figure 2b), the M.Peglia-Narnese Amerina EFS (E in Figure 2b) and the Eastern Sabina-Ernici EFS (F in Figure 2b).

West-Dipping Fault Alignments and Associated Earthquakes

The internal west-dipping seismogenic alignment (1 in Figure 2b) extends for nearly 150 km along strike. It can be divided into a northern half (1N), which consists of three seismogenic master faults (SMFs) (Rieti, Salto Valley and Velino-Magnola), and a southern half (1S), which consists of four SMFs (Campo Felice-Ovindoli, Fucino, M. Marsicano, Barrea boxes) (see Boncio et al., 2004 for the nomenclature of the seismogenic boxes). 1N is not well constrained both in terms of activity in the Holocene time and association with the earthquakes. There is no evidence of instrumental activity and the only relevant historical earthquake is the 1298 earthquake (I_0 VIII-IX MCS, Mw5.9), the source of which is highly uncertain and might be associated with the west-dipping border fault of the Rieti basin or alternatively with the east-dipping Narnese-Amerina segment of the EFS. Conversely, the southern half of the alignment (1S) caused the most destructive earthquake of central-northern Italy (Fucino 1915, M_s 7.0), associated with the Avezzano box, and a relevant instrumental seismic sequence, which activated the Barrea box in May 1984 (M_w 5.8) (Figure 5).

The intermediate west-dipping seismogenic alignment (2 in Figure 2b) extends for nearly 180 km and consists of 10 SMFs across the Umbria-Marche-Abruzzo Apennines (Gubbio, Gualdo T., Colfiorito, Cesi-Mt. Civitella, Nottoria-Preci, Cittareale, Montereale, Pizzoli, Paganica, and Middle Aterno Valley). The estimated depth of the seismogenic faults deepens southward along-strike from 6–7 km (Gubbio fault, at the northern termination of the alignment) to 14–15 km

(Middle Aterno fault, at the southern termination of the alignment) (Boncio et al., 2009). The entire alignment 2 shows a relevant instrumental and historical seismicity with moderate-to-large earthquakes (Figure 5, observed M_{max} 6.3). In the last thirty years, it has been affected by the earthquakes in Norcia 1979 (M_w 5.9), Gubbio 1984 (M_w5.6) and by a long seismic sequence, which went on from September 1997 to April 1998 and affected the Colfiorito, Sellano and Gualdo Tadino boxes (Colfiorito Sept. 26, 1997, M_w6.0; Sellano Oct. 12, 1997, M_w 5.2 and Oct. 14, 1997, M_w5.6; Gualdo T. April 3, 1998 M_w 5.1). In ancient times, the entire Gualdo T. box was activated during the July 1751 earthquake (I_0 X, M_W6.3) and possibly by the April 1747 earthquake (I_0 IX, M_w 5.9). The northern portion of the Nottoria-Preci box was activated in December 1328 (I_0 X, M_w 6.4,) and in August 1859 (I_0 VIII-IX, M_w 5.8) and the southern portion in May 1730 (I_0 VIII-IX, M_w 5.8; the entire box was activated from SE to NW by the highly destructive January 1703 earthquake (I_0 XI, M_w 6.8).

The Cittareale box can be associated with the November 1599 earthquake (I_0 VIII-IX, M_w 5.8); the Montereale and Pizzoli boxes were entirely activated by the highly destructive February 1703 earthquake (I_0 X, M_w 6.7); the Paganica box was recurrently activated by the 1461 (I_0 IX, M_w 5.9), 1762 (I_0 X, M_w 6.7), 1958 (M_w 5.0) earthquakes and by the recent L'Aquila 2009 event (I_0 IX, M_w 6.3) (Lavecchia et al., this vol.).

The external west-dipping seismogenic alignment (3 in Figure 2b) extends for nearly 160 km coinciding with the eastern breakaway of the intra-Apennine extensional system. It can be divided into a northern and a southern portion (3N and 3-S), which show homogeneous seismicity characters and a common inferred seismogenic depth of 14-15 km, based on both rheological and seismological data (Boncio et al., 2004, 2009). The 3N consists of three individual SMFs (M. Bove - M. Vettore, M. Gorzano - Campotosto and Gran Sasso). The northernmost and southernmost ones (Mt. Bove - M. Vettore and Gran Sasso, respectively) are silent since historical times, but show paleoseismological evidence of strong pre-historic earthquakes (Galadini and Galli, 2003; Galadini et al., 2003). The 1639 earthquake (I_0 X, M_w 6.3), which seriously damaged the Amatrice area, can be associated with the central fault (Mt. Gorzano - Campotosto). This earthquake suggests the activation of the northern portion of the structure, which has also been partly activated during the L'Aquila 2009 sequence (Lavecchia et al., this vol.).

The 3-S consists of four individual SMFs (Ofena, Morrone, Porrara, Pizzalto - Cinque Miglia). It is characterised by minor and rare instrumental activity, by some archeologically-inferred evidence of strong earthquakes for the Morrone

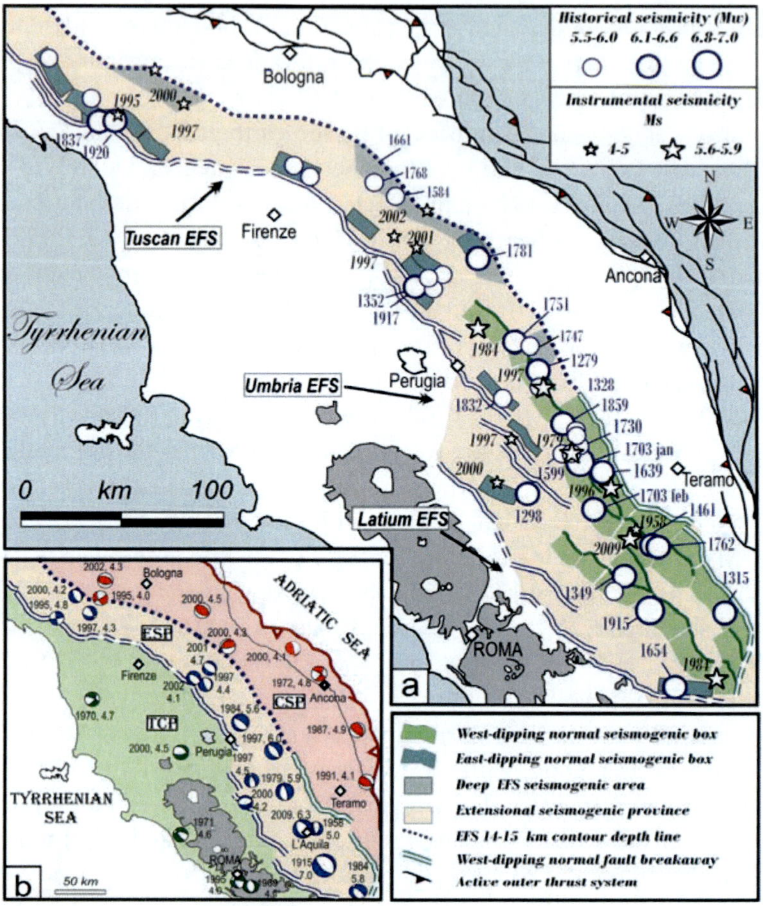

Figure 5. (a) Seismotectonic zonation of the central Italy active extensional province, with epicentres (circles) of the major historical earthquakes (from 1000 A.D. to 2004) (Working Group CPTI04, 2004) and instrumental earthquakes (stars). The western boundary of the province coincides with the surface trace of the EFS Tuscan, Umbrian and Latium branches, the eastern boundary with the surface projection of the ~14-15 km EFS contour depth line. The province is divided into an inner and outer domain by the surface trace of the west-dipping breakaway fault envelope. The rectangular shaped boxes represent the surface projection of individual seismogenic master faults drawn in Figure 2a (modified from Boncio et al., 2004). The east-dipping and west-dipping seismogenic fault alignments responsible for Mw≥5.5 earthquakes are identified in blue and green, respectively. (b) Major seismotectonic provinces in central Italy (modified from Lavecchia et al., 1994; Pace et al., 2006) with related available focal mechanisms for M≥4 earthquakes. Key: TCP = Thinned Crust Province, ESP = Extensional Seismogenic Province, CSP = Compressional Seismogenic Province.

box and paleoseismological evidence of strong pre-historic earthquakes on the Pizzalto-Cinque Miglia box (Galadini and Galli, 2000; D'Addezio et al., 2001).

East-Dipping Low-Angle Seismogenic Alignments

The northern portion of the Tuscan EFS (e.g., Garfagnana - North Apuane - Lunigiana seismogenic alignment, A in Figure 2b) extends for ~90 km with an average N130–140° strike (Brozzetti et al., 2007). It consists of five SMFs defined on a geological basis (Mulazzo, Olivola, north Apuane, Casciana-Sillicano and Bolognana-Gioviano) having a maximum seismogenic depth of 4 to 7 km (Figure 5). The alignment is characterised by the September 7, 1920 earthquake (I_0 IX-X, M_w 6.5), which damaged a large WNW-ESE striking area, from northern Garfagnana to southern Lunigiana, suggesting the possible activation of both the Casciana-Sillicano (northern Garfagnana) and the eastern portion of the North Apuane (southern Lunigiana) faults. The October 1995 strike-slip earthquake (M_w 4.9) may be related to the North Apuane fault together with the April 11, 1837 earthquake (I_0 IX-X, M_w 5.7) and a number of small-magnitude events ($4.5 \leq M \leq 5.2$) in 1767, 1902, 1928 and 1962. Computations of the ruptured areas on the basis of the estimated magnitudes of the associated earthquakes suggest that the entire Casciana-Sillicano fault plus the North Apuane transfer were almost entirely activated in historical times.

The southern portion of the Tuscan EFS (e.g., Mugello - Casentino - Alta Val Tiberina seismogenic alignment, C in Figure 2b) extends for ~95 km with an average N130° strike. It consists of three SMFs defined on a geological basis (Mugello, Poppi and Città di Castello faults) and having maximum seismogenic depth values of 6 to 8 km (Brozzetti et al., 2009 and reference therein). Two earthquakes with $M_w \geq 5.5$ can be associated with the Mugello source (June 13, 1542, I_0 IX, M_w 5.9 and June 29, 1919, I_0 IX, M_w 6.2) (Figure 5).

Three earthquakes with $M_w \geq 5.5$ can be associated with the Val Tiberina source (Oct. 18, 1389 I_0 IX, M_w 6.0; April 26, 1458 I_0 IX, M_w 5.9; Sept. 30, 1789 I_0 VIII-IX, M_w 5.8), plus a number of small earthquakes, including the 2001 earthquake (M_w 4.7), which is of special interest because its focal mechanism and aftershock sequence are consistent with the activation of a NE-dipping low-angle normal fault (~20° dip). No major historical or instrumental earthquakes can be associated with the Casentino western border fault, although the fault features are equivalent to those of the Mugello and Val Tiberina boundary faults. A moderate event probably linked with this structure is the Bibbiena November 1, 1504 earthquake (I_0 VI-VII, M_w5.0).

One should also note that the Tuscan and Umbria EFS could be seismically active also at mid-crust depth (Figures 5 and 6). The direct EFS association with a systematic microseismic activity down to depth values close to the brittle-ductile transition (nearly 14 km in the area) has been confirmed by instrumental data (Boncio et al., 1998; Piccinini et al., 2003; Chiaraluce et al., 2007) (Figure 4). The potential association with large historical earthquakes (e.g., Cagli June 3, 1781, I_0 IX-X, M_w 6.2) has also been pointed out, mainly based on the results of the fault parameters inversion from macroseismic data (de Nardis et al., 2009). In our opinion the deep Tuscan EFS segments might also be associated with some strong earthquakes, whose epicentral area is located eastward of the outermost west-dipping normal faults, but still within a crustal block with prevailing extensional mechanisms (Pondrelli et al., 2006), and far away from the active structures of the Adriatic thrust front. They are the September 10, 1584 (I_0 IX, M_w 6.0), the October 19, 1768 (I_0 IX, M_w 5.8) and the November 10, 1918 (I_0 VIII, M_w 5.8) earthquakes in the Romagna Apennines. The March 22, 1661 earthquake (I_0 IX, M_w 5.8), located more to the NE compared to the 1584, 1768 and 1918 events, is more difficult to interpret.

The Umbria EFS (e.g., the Umbertide - Umbria Valley – Bastardo - Leonessa basins alignment, D in Figure 2b) extends for ~150 km, with an average N140-150° strike. Evidence of moderate and minor seismogenic activity in association with the M. Santa Maria synthetic splay, at the northern termination of the alignment, and southward on the Umbria Valley faults, exists. The shallowest synthetic splays of the entire alignment are characterised by detachment depth values which never exceed ~5 km. Two historical earthquakes, which can be attributed to this alignment, are the December 25, 1352 (I_0 IX, M_w 6.0) and April 26, 1917 (I_0 IX, M_w 5.8) Monterchi events: they might have possibly activated the M.S.Maria Tiberina fault or the parallel Mt. Favalto fault. These two fault structures are the northward end of the alignment, which coincides with the Umbria EFS, and laterally overlap with the Città di Castello fault, which is the southernmost structure of the Tuscan EFS. The epicentral areas of at least eight earthquakes with $5.4 \leq M_w \leq 5.8$ are located in the Umbria Valley between Foligno and Spoleto, the largest known event being the January 13, 1832 Foligno earthquake (I_0 VIII-IX, M_w 5.8). During the last decades, the Umbria Valley has only experienced spread small-to-micro earthquakes.

The seismogenic role of the east-dipping fault systems, which belong to the Latium EFS branch alignment (E and F in Figure 2b), is only poorly defined. The December 01, 1298 (I_0 VIII-IX, M_w 5.9) event, with a doubtful epicentral location, and the Terni earthquakes of October 9, 1785 (I_0 VIII, M_w 5.5) and May 12, 1917 (I_0 VII-VIII, M_w 5.1) might be tentatively associated with the Narnese-

Amerina fault (E2 in Figure 2a). Alternatively, these events might be attributed to the SW-dipping Rieti and SSW-dipping Martani sources (antithetic EFS splays) (Boncio et al., 2000). In general, the Terni basin is characterised by moderate seismic activity, with local clusters of microseismicity and small-magnitude earthquakes. Near the southern end of the E alignment, the strong July 23, 1654 earthquake (I_0 IX-X, M_w 6.2) might be associated with the E-W striking Casalattico-Atina fault (F3 in Figure 2a), at the southern termination of the system. Alternatively, these events could be ascribed to the west-dipping Sora fault (Galadini and Galli, 2000).

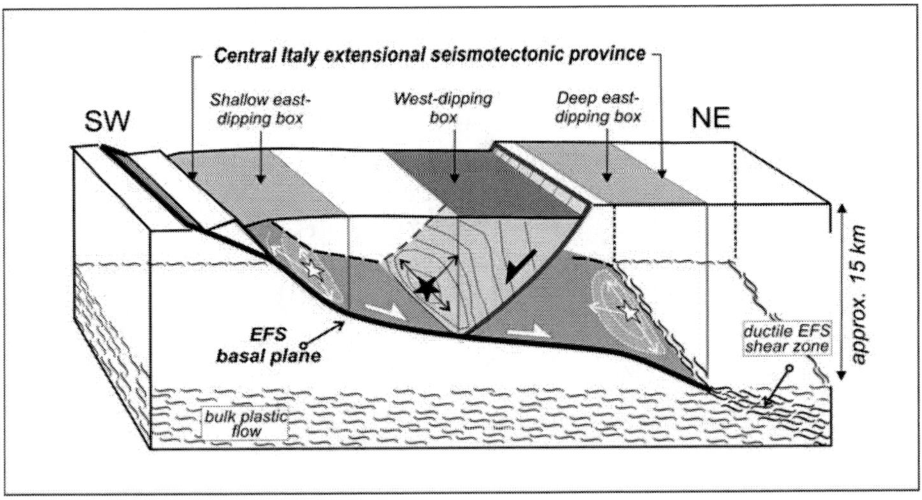

Figure 6. Schematic representation of the 3D geometry of the central Italy Extensional Seismotectonic Province and associated individual seismogenic faults and surface boxes. The sketch, not perfectly scaled, could be referred to a transect across the Umbria branch of the Etrurian Fault System, more or less at the latitude of Perugia.

The Multilayered Seismotectonic Zonation

The active extensional belt of central Italy, from northern Tuscany to southern Abruzzo (Figures 2a and b), is an area characterised by the availability of abundant detailed surface and sub-surface geological/geophysical data, together with good records of paleo- and historical earthquakes, as well as of instrumental major events and background seismicity. These data have been used to develop a multilayered seismotectonic zonation of the region, which consists of three overlapping layers: the Seismotectonic Province base layer, the Seismogenic Alignment (SA) intermediate layer and the Seismogenic Box (SB) top layer.

The *Extensional Seismotectonic Province* (SP) shown in Figures 5 and 6 coincides with the highly seismic domain of central Italy, which is undergoing an active SW-NE extension at the hangingwall of the EFS. The western limit of the province is the surface trace of the EFS system; the eastern limit is represented by the surface projection of the intersection line between the EFS-plane and the base of the brittle layer (~15 km on the basis of the depth-distribution of background seismicity and rheological considerations, Boncio et al., 2004 and references therein).

This large structural domain has a seismogenic thickness which increases from a few kilometres westward to ~15 km eastward, an average crustal thickness of 30 km in association with negative values of the Bouguer anomalies and heat flow values ranging from 40 to 70 mW/m^2. The depth-distribution of the seismicity and the thickness of the seismogenic layer within the entire province are controlled by two main factors: 1) the rheological stratification of the crust, which is a function of the crustal structure and heat flow, and 2) the presence of the Etrurian Fault System (EFS), which penetrates at low-angle within the upper and middle crust and acts as basal detachment of the other east- and west-dipping normal and normal-oblique extensional seismogenic structures (Boncio et al, 2000; Boncio and Lavecchia, 2000; Brozzetti et al., 2009).

The province can be divided into a western highly seismogenic domain and an eastern slightly seismogenic domain, separated by the surface breakaway zone of the west-dipping extensional system. The synthetic and antithetic faults within the EFS hangingwall divide this extending rock volume into crustal blocks with different seismogenic characters.

The *Seismogenic Alignment* (SA) layer consists of rectangle-shaped long and narrow polygons, which are the surface projection of the seismogenic volumes at the hangingwall of major aligned seismogenic fault systems responsible for large earthquakes ($M_w \geq 5.5$). Most of the large earthquakes are associated with the west-dipping alignments, which prevail in the south-eastern area of the province, but a remarkable component of the seismic release is also associated with the east-dipping sources, which prevail in the north-western side of the province. The long-term geological data and the present seismicity consistently show that the importance, in terms of accommodated offset of the east-dipping alignment, progressively decreases southward, whereas that of the west-dipping system decreases in the opposite sense, e.g. northward.

The *Seismogenic Box* (*SB*) layer shows a pattern which is strongly influenced by the segmentation of the fault alignments. In the case of the AFS west-dipping sources, such segmentation is constrained by surface geology data, which can be used to identify the lateral fault terminations. In the case of the EFS east-dipping sources, the definition of the single segments is more difficult and often based on the along-strike extent of the hangingwall basins. The maximum damage produced by the strong historical earthquakes ($M_w \geq 5.5$) associated with the boxes is recorded within the boundaries of the boxes. Therefore, also the macroseismic epicentre is located within, or close enough to, the corresponding box. Sources that have not been activated since ancient times can be identified thanks to geological and/or paleoseimological evidences.

To date, we have identified and parameterised 37 seismogenic boxes, 23 corresponding with west-dipping faults and 14 with east-dipping faults (Figure 5). Nine boxes are not associated with any known historical or instrumental earthquake. Hence the importance of their identification, from the viewpoint of the seismic hazard assessment implications.

CONCLUSION

At the end of this chapter, two key points should be highlighted. They concern: (A) the methodological approach that has been followed in constructing the seismotectonic zonation, which is multidisciplinary and strongly based on structural-kinematic data and (B) the collected evidence supporting the seismogenic low-angle normal faulting in central Italy.

A) - The integration of surface geology data with seismological and subsurface structural data has favoured the identification of the active fault pattern of the Late Pliocene-Quaternary extensional belt in central Italy, and the definition of the 3D geometry and seismogenic role of the major individual sources, as well as the structural style of the active deformation at regional scale. The implemented seismological data refer to both, historical and high quality instrumental earthquakes. Integrating the two is essential. The former is the widest and most complete source of information on the seismic activity of a region, but cannot give detailed information on the geometry and the kinematics of the seismogenic processes. The latter gives information on the size, shape, depth, kinematics of the seismogenic processes, but is highly heterogeneous and certainly not representative of the entire seismic activity. Geological-structural, morphotectonic and paleoseismologic data are essential for seismotectonic. They are of paramount importance to define the geometry, kinematics, timing and segmentation pattern of the potentially seismogenic faults. Furthermore, a 3D structural-seismotectonic approach, such as the one here presented, is suitable for a seismotectonic zonation capable of improving the seismic hazard assessments. Certainly, the 3D geometric reconstruction of seismogenic master faults can be a strong constraint to delimit the zones where the largest earthquakes are expected to occur. Unfortunately, the definition of seismogenic boxes is not always possible. This is especially true for seismic regions where the geometry of the master faults cannot be detailed by means of surface observations (e.g. blind or buried faults). Nevertheless, a multidisciplinary study based on the integrated analysis of the structural-geological, geophysical, rheological and seismological data can favour the creation of a plausible model for active tectonics and constrain the surface and deep extent of zones that are kinematically and seismotectonically homogenous.

B) On the basis of the surface and deep geology of the Etrurian Fault System (EFS), constrained by geology, seismic reflection data and distribution of instrumental and historical seismicity, the east-dipping low-angle normal faults (e.g., LANFs) can be identified as very important structures in controlling the active extensional process and the spacing and distribution of homogenous seismotectonic domains and volumes in central Italy. In a few cases, these Italian LANFs appear to be also responsible for some strong earthquakes (Brozzetti et al., 2009; de Nardis et al., 2009). These assumptions are mainly based on the following:

1) - A meaningful part of the horizontal deformation of the EFS east-dipping faults occurred during the Late Pliocene-Quaternary extensional phase, with offsets in the order of 5-7 km, and at least in Early Pleistocene-Holocene times they primarily controlled the syn-sedimentary growth of continental deposits.

2) - The main branches of the EFS extend along-strike for dozens of kilometres delimiting long grabens or systems of basins, whereas the west-dipping faults are limited to single basins or sub-basins (Figure 2a). In general, the west-dipping faults (antithetic faults) are hierarchically secondary structures compared to the EFS (synthetic faults), though their dimensions might be locally important in from the point of view of the seismogenic potential (i.e. able to generate large earthquakes);

3) - Along most of the Tuscan EFS, the east-dipping faults extend into the mid-crust and cut through basement units, whereas the antithetic west-dipping faults have limited down-dip length (4–5 km) and cut mostly through low-strength turbiditic deposits; these latter conditions are mechanically less favourable for a strong seismogenic potential.

4) - The eastward-deepening of the EFS system, down to ~15 km beneath the axis of the Apennines and its seismogenic role at mid-crustal depths, possibly on steeper down-dip segments (e.g., 25°–30°), might explain the occurrence of moderate-to-large earthquakes also in areas which are located eastward of the outermost alignment of outcropping AFS west-dipping normal faults (AFS breakaway west-dipping fault zone in Figure 2a).

In conclusion, within the general framework of the central Italy extensional province, the low-angle east-dipping faults of the EFS system play a primary role in controlling the shape and size of the deforming seismogenic volumes. The possibility of the association of a low-angle extensional plane also with large historical earthquakes is relevant because of the worldwide rarity of such evidence (e.g., Wernicke, 1995).

References

Argnani, A.; Barbacini, G.; Bernini, M.; Camurri, F.; Ghielmi, M.; Papani, G.; Rizzini, F.; Rogledi, S.; Torelli, L. *Quat. Int.* 2003, 101, 13-26.

Barchi, M.R.; Amato, A.; Cippitelli, G.; Merlini, S.; Montone, P. *Boll. Soc. Geol. Ital.* 2007, Spec. Iss. 7, 47-56.

Barchi, M.R.; De Feyter, A.; Magnani, M.B.; Minelli, G.; Pialli, G.; Sotera, B.M. *Mem. Soc. Geol. It.* 1998, 52, 527–538.

Barchi, M.R.; Paolacci, S.; Pauselli, C.; Pialli, G.; Merlini, S. *Boll. Soc. Geol. Ital.* 1999, 617-625.

Basili, R.; Valensise, G.; Vannoli, P.; Burrato, P.; Fracassi, U.; Mariano, S.; Tiberti, M.; Boschi, E. *Tectonophysics.* 2008, doi:10.1016/j.tecto.2007. 04. 014.

Benvenuti, M. *Giorn. Geologia.* 1997, 59, 91–111.

Benvenuti, M. *Sediment. Geol..* 2003, 157,197-234.

Bernini, M.; Papani, G. *Boll. Soc. Geol. Ital.* 2002, 121, 313–341.

Boncio, P.; Brozzetti, F.; Lavecchia, G. *Tectonics.* 2000, 19, 1038-1055.

Boncio, P.; Brozzetti, F.; Ponziani, F.; Barchi, M.; Lavecchia, G.; Pialli, G. *Mem. Soc. Geol. It.* 1998, 52, 539-555.

Boncio, P.; Lavecchia, G. *J. Geodyn.* 2000, 29, 233-244.

Boncio, P.; Lavecchia, G.; Pace, B. *J. Seismol.* 2004, 8, 407-425.

Boncio, P.; Tinari, D.P.; Lavecchia, G.; Visini, F.; Milana, G. *Boll. Soc. Geol. It. (Ital. J.Geosci.)* 2009, *128*, 367-380.

Brozzetti, F. Studi Geol. Camerti. 1995, Vol. *Spec.* 1995/1, 105-119.

Brozzetti, F.; Boncio, P.; Lavecchia, G.; Pace, B. *Tectonophysics.* 2009, 463, 31-46.

Brozzetti, F.; Boncio, P.; Tinari, D.P.; Di Naccio, D.; Torelli, L. Rend. *Soc. Geol. It.* 2007, 4, 164-165.

Chiaraluce, L.; Amato, A.; Chiarabba, C.; Selvaggi, G.; Di Bona, M.; Piccinini, D.; Deschamps, A.; Margheriti, L.; Courboulex, F.; Ripepe, M. *Bull. Seism. Soc. Am.*. 2004, 94, 99-116.

Chiaraluce, L.; Barchi, M.R.; Collettini, C.; Mirabella, F.; Pucci, S. *Tectonics.* 2005, doi:10.1029/2004TC001627.

Chiaraluce, L.; Chiarabba, C.; Collettini, C.; Piccinini, D.; Cocco, M. *J. Geophys. Res.* 2007, 112, doi:10.1029/2007JB005015.

Ciaccio, M.; Barchi, M.; Chiarabba, C.; Mirabella, F.; Stucchi, E. *Tectonophysics.* 2005, 406, 233-247.

Collettini, C.; Barchi, M. *Tectonophysics.* 2002, 359, 97-115.

D'Agostino, N.; Mantenuto, S.; D'Anastasio, E.; Avallone, A.; Barchi, M.; Collettini, C.; Radicioni, F.; Stoppini, A.; Castellini, G. *Tectonophysics.* 2008, doi:10.1016/j.tecto.2008.09.033.

D'Addezio, G.; Masana, E.; Pantosti, D. *J. Seismol.* 2001, 5, 181-205.

de Nardis,, R.; Lavecchia, G.; Pace, B.; Visini, F. *Geophys. J. Int.* 2009, in press.

Galadini, F.; Galli, P. *Nat. Hazards.* 2000, 22, 225-270.

Galadini, F.; Galli, P. *Ann. Geophys.* 2003, 46, 815-836.

Galadini, F.; Galli, P.; Moro, M. *Ann. Geophys.* 2003, 46, 793-813.

Hunstad, I.; Selvaggi, G.; D'Agostino, N.; England, P.; Clarke, P.; Pierozzi, M. *Geophys. Res. Lett.* 2003, doi:10.1029/2002GL016447.

Lavecchia, G.; Boncio, N.; Creati, N. *J. Geodyn.* 2003, 36, 79-94.

Lavecchia, G.; Boncio, P.; Brozzetti, F.; Stucchi, M.; Leschiutta, I. *Mem. Soc. Geol. It.* 2002, 1, 881-890.

Lavecchia, G.; Brozzetti, F.; Barchi, M.; Keller, J.V.A.; Menichetti, M. *Bull. Seism. Soc. Am.* 1994, 106, 1107-1120.

Meletti, C.; Galadini, F.; Valensise, G.; Stucchi, M.; Basili, R.; Barba, S.; Vannucci, G.; Boschi, E. *Tectonophysics.* 2008, 450, 85-108.

Meletti, C.; Patacca, E.; Scandone, P *Pure Appl. Geophys.* 2000, 157, 11-35.

Montone, P.; Mariucci, M.T.; Pondrelli, S.; Amato, A. *J. Geophys. Res.* 2004, doi:10.1029/2003JB002703.

Pace, B.; Boncio, P.; Lavecchia, G. *Tectonophysics.* 2002, 350, 237-254.

Pace, B.; Peruzza, L.; Lavecchia, G.; Boncio, P. *Bull. Seism. Soc. Am.* 2006, 96,107-132.

Piccinini, D.; Cattaneo, M.; Chiarabba, C.; Chiaraluce, L.; De Martin, M.; Di Bona, M.; Moretti, M.; Selvaggi, G.; Augliera, P.; Spallarossa, D., Ferretti,

G., Michelini, A.; Govoni, A.; Di Bartolomeo, P.; Romanelli, M.; Fabbri, J. *Ann. Geophys.* 2003, 46, 1315-1324.

Pondrelli, S.; Salinbeni, S.; Ekström, G.; Morelli, A.; Gasperini, P.; Vannucci, G. *Phys. Earth Planet. Inter.* 2006, 159, 286-303.

Valensise, G.; Pantosti, D. *Ann. Geophys.* 2001, 44, 1-180

Wernicke, B. *J. Geophys. Res.* 1995, 100, 20159–20174.

Working Group CPTI (2004). Catalogo Parametrico dei Terremoti Italiani, vers. 2004 (CPTI04), INGV, Bologna. http://emidius.mi.ingv.it/CPTI04/.

In: Recent Progress on Earthquake Geology ISBN: 978-1-60876-147-0
Editor: Pierpaolo Guarnieri, pp. 189-247

Chapter 8

THE RESULTS OF SUBTERRANEAN ELECTRIC MEASUREMENTS ON KAMCHATKA AS GLOBAL EFFECTS OF PROTON TECTOGENESIS: DAMAGING EARTHQUAKES IN INDONESIA AND CHINA

V.S. Bobrovskiy*

Subterranean Electric Technologies Innovation Foundation,
Petropavlovsk-Kamchatskiy 683024, Russia

ABSTRACT

Measurements of subterranean electromotive force (SEMF) of alternating and direct currents show abrupt changes in the fine structure of proton migration before earthquakes. This chapter reviews convincing evidence of worldwide relationships between earthquakes and proton electric (PE) measurements taken a world away from a seismic event epicenter. In turn, this places proton tectogenesis at the primary position as the cause of the sudden onset of an earthquake. Proton tectogenesis, having an electric nature, is a worldwide phenomenon and can be described in terms of electrical current networks.

---

* Address: Office 16 Bld. 15/2 Prospekt 50 let Oktyabrya, Petropavlovsk-Kamchatskiy 683024, Russia. E-mail: Vad.Bobrovskiy@cosmetecor.org; vadokoem@gmail.com

Then, we report the description of subterranean PE measurement instruments in Petropavlovsk-Kamchatskiy. The system is used to research the fine structure of proton migration at the tectonosphere-atmosphere interface. The results of PE measurements in Petropavlovsk-Kamchatskiy have shown correlations with the tsunamigenic earthquake in Indonesia (December 26, 2004) and the damaging earthquake in China (May 12, 2008).

This chapter reviews the advance of the description of subterranean PE processes in terms of electrical current networks because there are difficulties with electromagnetic interpretation of observations of low-frequency magnetic field variations.

Keywords: proton tectogenesis, earthquake, Kamchatka, proton electric measurements, ULF magnetic field variations

INTRODUCTION

Earthquake geology has developed in two directions.

Plate tectonics is one of the directions taken by research. It was considered that theoretical propositions could be generalized based on all of the processes in the Earth's interior. In addition, these theoretical propositions claimed to be exhaustive for an explanation of all of the properties, nature and causes of earthquakes. But these efforts went off like a damp firecracker. The problem appeared much more difficult than previously thought and required some simplifications. At times, theoretical research was contrary to experimental investigation. Compound and incomprehensible phenomena observed during the onset of an earthquake were simply mapped out.

Proton tectogenesis (Larin, 1980; Larin, 1993) is the basis of another research direction. Such concepts as block drifts, rock movements in joint fissures, etc. explain the reasons for the slow phase at the beginning of an earthquake. But proton migration from Earth's core to circumterrestrial space (Larin, 1980; Larin, 1993) is the reason for the rapid phase of the onset of an earthquake. Under external factors, pulses of protons penetrate from the Earth's core through all of the Earth's envelopes. When, penetrated by protons, the Earth's interior volume is well developed and prepared, we observe the onset of an earthquake.

The worldwide scale phenomenon made clear the relationship of earthquake epicenters from thousands of kilometers' distance. For the first time, Båth (1966) discovered a seismic-global effect when studying the gigantic earthquake near the Aleutian Islands and Alaska. The cumulation of a global seismic-resource in the zone of a gigantic earthquake and its aftershocks has caused a decrease of seismic resource in the remaining parts of Earth's subterranean layers (Båth, 1966). The velocity of cumulation excludes the possibility of its explanation on the basis of mechanical background.

B. Gutenberg's concept of cosmic-meteorologic-seismic correlations (Gutenberg, 1927) has become the key for the interpretation of seismic-global effects. This concept was consigned to oblivion for a long time. The "cosmic phenomena" term in paragraph §11 (Gutenberg, 1927) was related to sunspots. Magnetic fields of sunspots were considered surface events of electric currents flowing in deep layers of the Sun. The "meteorological variables" term was related with structures appearing as a result of electric dipole (water molecules and its dimers, trimers, tetramers and multimers) condensation, as it was believed, on charged particles of dust in the Earth's atmosphere. Thereby, "cosmic-meteorologic-seismic parallelism" (Gutenberg, 1927, §11) is based on the study of electricity and matter.

After H. Alfven (1981) suggested an electric-current network approach to describe interactions of solar wind, the Earth's magnetosphere and ionosphere, it became clear that the same electric-current network approach could be used for the Earth's interior. Hereby, an explanation was found for how positive elementary charges (hydrogen atomic nuclei, i.e., protons) supply rapid global relationships of earthquakes' epicenters through the Earth's core. As another consequence of this approach, it is necessary to follow up electric phenomena prior to, during, and after earthquakes.

A subterranean PE measurement technique for registration of fine structure of proton migration through heterogeneities at the tectonosphere-atmosphere interface was devised and applied practically on Kamchatka. These PE measurements corroborated the natural occurrence of a global seismic effect. Also, these PE measurements made it possible to record abrupt changes in the fine structure of proton migration before earthquakes. PE measurements also corroborate the appropriateness of returning from electrotelluric field conception to telluric currents, one being in subterranean electric current networks.

We recall that the only one standard in classical electricity is the normal cell, preserving fixed force, and moving electricity. Electromotive force (EMF, volt) translates into other electrical units with the help of rules of thumb (by Coulomb, Ampere, Biot-Savart-Laplace, Ohm, Joule, Kirchhoff, etc.). In developing conceptions, it is possible to translate any parameter of any nonstationary process involving charges in subterranean, overground and nonterrestrial current networks into EMF quantity with the help of such rules of thumb.

The task of research is to review convincing evidence of worldwide relationships of earthquakes and PE measurements taken a world away from an incipient seismic event epicenter. This is the relationship that places proton tectogenesis (Larin, 1980; Larin, 1993) in the primary position as the cause of rapid earthquake inception. Proton tectogenesis, having an electric nature, is global in nature and can be described in terms of electric current networks.

Taking PE measurements on Kamchatka makes it possible to follow up on the abrupt changes in the subterranean global current network and provide a warning of the beginning of the rapid phase of earthquake inception. An example of the success of such an approach was reviewed at Erice International School of Geophysics' 26th workshop on "Earthquake and Shaking Probabilities: Helping Society to Make the Right Choice" (October 18–24, 2006, Sicily, Italy). The presentation reviewed implementation of short-term forecast of a strong earthquake on Kamchatka in the winter–spring of 2006. Forecasting and its illustrative materials were presented at the conference "Decrease of Seismic Risks in Kamchatka-Kuril Zone." The conference was organized by the Ministry of the Russian Federation for Civil Defense and took place on December 7, 2005 in Petropavlovsk-Kamchatsky. After this, forecast and its enlarged illustrative materials were published in All-Russian institute for Scientific and Technical Information Russian Academy of Science (Bobrovskiy, 2005). The seismic forecast was realized by a strong M7.6 earthquake (April 20, 2006, UT) in the north of the Kamchatka peninsula. Data here and cited below are taken from the U.S. Geological Survey [http://neic.usgs.gov/neis/epic/].

This chapter has reviewed variations in subterranean PE signals measured on Kamchatka. The chapter deals with abrupt changes of fine structure of proton migration before the tsunamigenic earthquake in Indonesia (December 26, 2004) and also before the damaging earthquake in the Sichuan province in southwestern China (May 12, 2008).

METHODS

It is proposed in the framework of current-network conception that nonstationary processes occur in the subterranean current networks. The processes are able to occur with intensity difference everywhere in the Earth's crust where protons approach the surface with a different in surface density. Subterranean nonstationary electric processes were investigated by measuring lines whose ends had lead electrodes buried in bore pits at a depth of some meters. Measuring line lengths varied from kilometers to tens of meters. Usually, a couple of measuring lines, extended along magnetic meridian and parallel, are used for world-wide telluric current component recognition. Such "crosses" are being used up to now to measure telluric currents to recognize nonstationary processes occurring before earthquakes in seismic-hazardous areas (Greece [Varotsos et al., 1984; Varotsos et al., 2006] and Japan [Uyeda et al., 2002]). But it is impossible to gather information on the fine structure of subterranean PE processes and telluric currents occurring in soils at the tectonosphere–atmosphere interface with the help of such instruments. It is necessary to use multielectrode systems to gather information on fine structure of subterranean PE processes.

As it is well known, proton conductivity of ordered water structure is much greater than proton conductivity of water fluid and its solutions (Zaretskiy et al., 1988; Khusnatdinov et al., 1987). The Earth's crust under the ocean floor has comparatively little thickness in relation to the Earth's crust under land. Therefore, the most intense proton migration in the hydrosphere is through channels located under the ocean (sea) floor. From channels approaching the underwater surface of the Earth's crust (ocean floor), protons move in an ordered structure. In other words, protons move within the limits of double electric layer on the partition of the solid and liquid phase (ocean floor water). In the framework of the current-network nature of earthquakes, nonstationary processes in the current networks could be the most distinctly measured in the tectogene covering the ocean (sea) floor and coast. At the same time, proton output occurs mainly close to the shoreline.

Thus, it is possible to take the most informative subterranean PE measurements with the help of electrodes digging into the bore pit being close to shoreline to recognize nonstationary processes occurring in the subterranean current networks. It must be considered that if the measuring station is located on the land far from the shoreline, the greatest surface density of channels is located in the least solid rocks. Less solid rocks are greatly influenced by flowing waters; therefore, it is necessary to locate the measuring bore pit as closely as possible to the bank of flowing waters.

Three bore pits of Station No. 1 (S1) of the Distant Experts School Polygon (DESP, 53°03′10.43″N, 158°39′10.86″E) are situated on the pitch of hill "Gorizont," about 150 meters from a stream, running toward the coast of Avacha Bay in Petropavlovsk-Kamchatskiy. To recognize nonstationary processes occurring in subterranean current networks, it uses systems of six electrodes digging into the southwestern (SW) bore pit, about 2 meters deep, and the central (C) bore pit has the same depth. There are two electrodes in the northeastern (NE) bore pit, about 1.5 meters deep. The NE bore pit and SW bore pit are located about 10 meters from the C bore pit. Electrodes 0.25*0.25*0.003 m^3 in size are layered one above another in the bore pits. Electrodes are iron because it is swelling comparatively much with other metals (up 36% [Alefeld et al., 1978]) in fluid hydrogen. It's possible only with much protons "affinity" with iron atoms. There are six sequence electrodes in sequence No.1–No.6 top-down in SW-bore pit and C bore pit. Depth difference between two sequence electrodes is not up 0.3 m. There are only two electrodes in NE bore pit, with a depth difference of not more than 0.5 m. Such small distances between electrodes make it possible to take information on fine structure of subterranean PE processes, occurring at the surface terrain.

On being interpreted the results of subterranean PE measurements the idea is that every couple of electrodes in bore pits and terrain volume between them are represented as subvertical electrode-electrode PE-two-pole network. Water and heating pipes joint with neutral main of household three-phase network is called structure. Every structure-electrode PE-two-pole network is confronted with fuzzy terrain volume. Every two electrodes in the NE bore pit (bottom electrode No. 2—NE2) and in C bore pit (C1, C2…C6) are called subhorizontal subterranean PE-two-pole network. Every subhorizontal electrode-electrode subterranean PE-two-pole network (NE2C1, NE2C2,…NE2C6) is confronted with fuzzy terrain volume. The general number of PE-two-pole network for measurements of PE parameters is 29. PE characteristics are measured with the help of multimeter M890G (by Precision MASTECH Enterprises Company, Hong Kong). Subterranean electromotive forces (SEMF) of direct (SEMF=, =ε) and alternating (low-frequency) currents (SEMF~, ~ε) are measured. Thus, the general number of measuring PE characteristics (=ε and ~ε) are 58 at S1-DESP-PK. Our subterranean PE measurements cover narrow frequency band 30 Hz–500 Hz, with a time resolution varying from 3 to 6 hours and attached to apparent solar time (AST) in Petropavlovsk-Kamchatskiy.

Daily measurements of subterranean PE parameters were begun on April 4, 2001 at S1-DESP-PK, i.e., approximately in peak of Wolf numbers. Since that time, the most damaging earthquake was in Indonesia on December 26, 2004 with magnitude M9, which initiated the highest destructive wave.

The longest continuous subterranean PE measurement series was begun in 1989 on Kamchatka. Petropavlovsk-Kamchatskiy is the most suitable place to establish a centre for a global network of subterranean PE monitoring of dangerous tectonic regions. The set-up costs will be reasonably moderate, considering the lowcost of iron electrodes and minimal digging efforts to bury electrodes into the bore pits. Analog-digital (AD) conversion of SEMF and measurement-processing equipment are also inexpensive; the most simple AD-converters and home personal computers will be used. The reasonably quick connection to an expert system of operative earthquakes forecast of networked stations of PE monitoring in different dangerous tectonic regions will have to cover the expenses for expert system engineering.

For comparison, some results of using of ultra-low-frequency (ULF) magnetic field variation stations (Hattori et al., 2004; Hayakawa et al., 2007) are quoted. Magnetic field sensors and observation systems used for seismic direction finding are situated in Japan. A number of circumstances, however, make seismic direction finding projects unsuitable for operative earthquake forecasting. Analyzing results of ULF magnetic field variation for operative earthquake forecasting is difficult—the rather limited information of the subterranean currents is being gathered by ULF magnetic field sensors. This is the main circumstance.

To use ULF magnetic field variations for seismic direction finding, another circumstance presents difficulties. The interpretation of results of ULF magnetic field variations taken with the help of the magnetic field sensors is not representable in the framework of the electromagnetic wave theory. Let us discuss the unrepresentability of ULF magnetic field variation precursors by way of example of the Loma Prieta earthquake on October 17, 1989 at 04:15.24 UT, a moderately large M_s 7.1. Its epicenter (φ37.039°N, λ121.879°W) was located near Mt. Loma Prieta in the California Coast Range. The earthquake occurred "suddenly and without foreshock activity" (Fraser-Smith et al., 1990). The paper (Fraser-Smith et al., 1990) reviewed information on ULF magnetic field variations that were interpreted in the framework of the electromagnetic wave theory. Low limit of frequency (ν) ULF band, which covered measurements (Fraser-Smith et al., 1990), is 0.01 Hz (period T=100 s, electromagnetic wavelength $\lambda = c/\nu \cong 3*10^7$ km, i.e., 30 million km). One electromagnetic noise monitoring system (ULF system; frequencies 0.01 Hz–10 Hz) was located at Corralitos, California (φ37.015°N, λ121.806°W), only 7 km from the epicenter, and the other (the extremely low/very low frequency [ELF/VLF] system; frequencies 10 Hz–32 kHz) was located on the Stanford campus (φ37.43°N, λ122.18°W), about 52 km from the epicenter (Fraser-Smith et al., 1990). Review of variations of the Corralitos 0.01 Hz magnetic field measurements, the closest to the Loma Prieta earthquake epicenter, during October 1989 (Fraser-Smith et al., 1990, Figure 3) discovered an increase to an exceptionally large level of activity starting approximately three hours before the earthquake. But review of the Stanford, the furthest to the Loma Prieta earthquake epicenter, shows that the ELF/VLF data did not appear to show precursor activity (Fraser-Smith et al., 1990), despite both electromagnetic monitoring systems having frequencies band edge on 10 Hz. It is concluded due to the above circumstances that the interpretation of results of low-frequency magnetic field measurements must not be based on the electromagnetic wave theory.

Let us make some calculations in the framework of the electromagnetic wave theory. The first station does indeed show that $L_1 = (7/30000000) \cong 2*10^{-7}$ as electromagnetic wave fraction; in another one, $L_2 = (52/30000000) \cong 2*10^{-6}$. It is generally known that, to interpret measuring signals in the framework of the electromagnetic wave theory, interferometers can be used with antennas and arms (L) equivalent to $n\lambda$, λ, $\lambda/2$, $\lambda/4$, …, and just can't seem to be $\lambda/10^6$ or less. This implies, in the framework of the electromagnetic wave theory, that the sensors of both stations (Fraser-Smith et al., 1990) are equivalent to a noticeable fraction of $\lambda \cong 3*10^7$ km. But being located on the interferometer arm (at a distance between these stations) which is an infinitesimal fraction of electromagnetic wave $L_{12} \cong (45/3*10^7) \cong 1.5*10^{-6}$, the sensors can't record such strong differentiating signals in the frequency bands having the one band edge. Obviously, attempting interferometer interpretation of measurements is a false assumption for such closely (in electromagnetic wave fractions) located low-frequency magnetic field stations (Fraser-Smith et al., 1990).

Being unaccountable from the position of electromagnetic wave theory, anomalous features of low-frequency magnetic field signals from Corralitos station (only 7 km from the Loma Prieta earthquake epicenter) and Stanford campus station (about 52 km from the Loma Prieta earthquake epicenter) (Fraser-Smith et al., 1990) have a well-exhaustive explanation in the framework of the proton tectogenesis model (Larin, 1980; Larin, 1993). In its framework, the intensity of "proton blowing" of faults and craters has been increased in the phase of preparation of focus of a seismic event and its inception. But the intensity of "proton blowing" of rock faults or degassing body is not increased so much. Thus, anomalous differences of low-frequency magnetic field signals from Corralitos station (only 7 km from fault) and Stanford campus station (about 52 km from fault) are due to differences of proton permeability of rocks beneath the sensors in the framework of the proton tectogenesis model (Larin, 1980; Larin, 1993). Such differences of proton permeability of Earth's interior cause the subvertical proton currents to differentiate through the low-frequency magnetic field sensors (Fraser-Smith et al., 1990) involving the iron cylinders covered by wire. Low-frequency magnetic field measurements interpreting as a measurement of a magnetic component of electromagnetic wave, indeed, carry the precursor information of proton bursts through the iron cylinders and wires, protons having been taken much by transiton metals (Fe, Ni, etc.).

RESULTS

The results of measurements of SEMF~ (subterranean PE signals of alternating currents) at S1-DESP-PK have been reviewed. SEMF~ variations were selected from 58, the most significant to recognize subterranean PE relationships of Kamchatka and epicenters of damaging earthquakes in Indonesia (December 26, 2004, magnitude M9) and China (May 12, 2008, magnitude M7.9).

Figure 1 is a column diagram of the largest daily signals of SEMF~ taken in C23~ channel from November 20 to December 31, 2004. Most of the fluctuations in the November indices are typical of those observed during earlier months and can be considered normal. However, over the interval December 1–3, an anomalous nonstationary process began to appear in the data for the SEMF~C23, and it dropped down until it was back comparatively to the previous data over the interval December 4–15. Then the main nonstationary process began and it dropped abruptly around December 16. It was not until the drop down that the tsunamigenic earthquake with magnitude M9 in Indonesia (December 26, 2004) occurred after about 11 days.

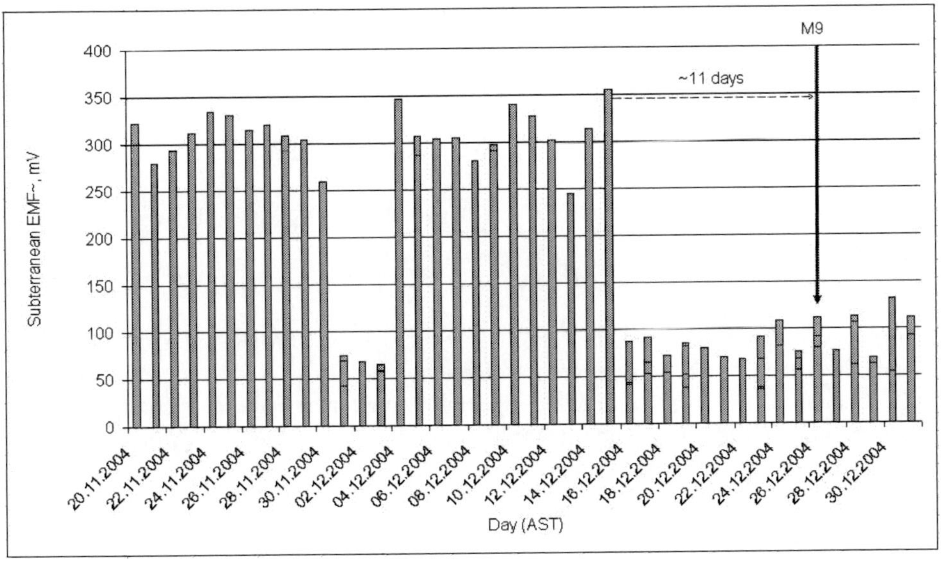

Figure 1. Column diagram of the largest daily signals of SEMF~ being taken in C23~ channel over the interval 2004 November 20 to December 31, 2004. AST – apparent solar time in Petropavlovsk-Kamchatskiy; EMF~ – electromotive force of alternating currents; M9 – Indonesian earthquake magnitude (December 26, 2004); ~11 days – interval of PE signals abrupt jump downward before earthquake M9.

Figure 2 is a column diagram of the largest daily signals of SEMF~ taken in C45~ channel over the interval November 20 to December 31, 2004. Most of the fluctuations in the November indices are typical of those observed during earlier months and can be considered normal. However, over the interval December 1–3, an anomalous nonstationary process began to appear in the data for the SEMF~C45 and it dropped down until it was back a comparatively to the previous data over the interval December 04–15. Then the main nonstationary process began and it abruptly dropped down around December 16. It was not until the jump downward that the tsunamigenic earthquake with magnitude M9 in Indonesia (December 26, 2004) occurred after about 11 days.

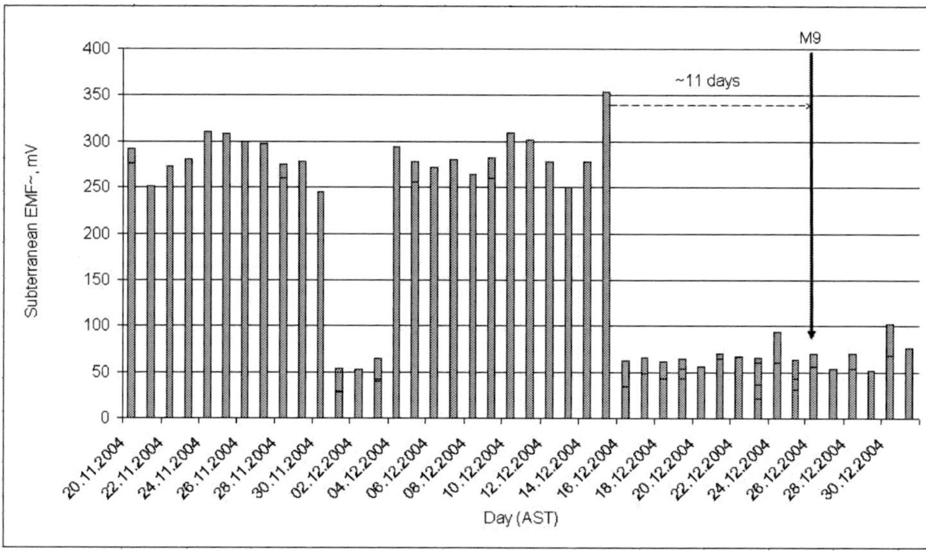

Figure 2. Column diagram of the largest daily signals of SEMF~ being taken in C45~ channel over the interval November 20 to December 31, 2004. AST – apparent solar time in Petropavlovsk-Kamchatskiy, EMF~ – electromotive force of alternating currents, M9 – Indonesian earthquake magnitude (December 26, 2004), ~11 days – days interval of PE signals abrupt jump downward before earthquake M9.

Figure 3 is a column diagram of the largest daily signals of SEMF~ being taken in CO1~ channel over the interval November 20 to December 31, 2004. Most of the fluctuations in the November indices are typical of those observed during earlier months and can be considered normal. However, over the interval December 1–3, an anomalous nonstationary process began to appear in the data for the SEMF~CO1 and it dropped down until it was back comparatively to the previous data over the interval December 4–15. Then the main nonstationary process began and it abruptly dropped down around December 16. It was not until the jump downward that the tsunamigenic earthquake with magnitude M9 in Indonesia (December 26, 2004) occurred, after about 11 days.

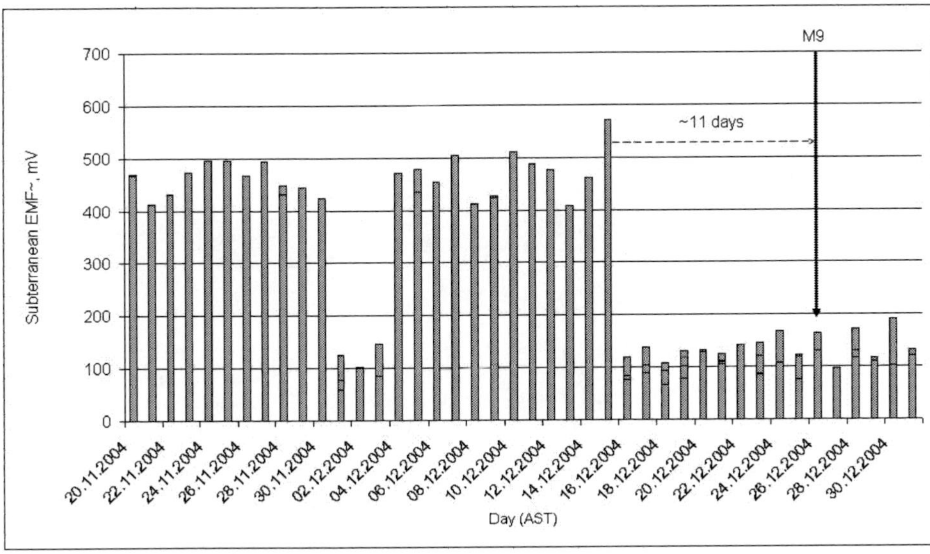

Figure 3. Column diagram of the largest daily signals of SEMF~ being taken in CO1~ channel over the interval November 20 to December 31, 2004. AST – apparent solar time in Petropavlovsk-Kamchatskiy, EMF~ – electromotive force of alternating currents, M9 – Indonesian earthquake magnitude (December 26, 2004), ~11 days – days interval of PE signals abrupt jump downward before earthquake M9.

Figure 4 is a column diagram of the largest daily signals of SEMF~ being taken in CO2~ channel over the interval November 20 to December 31, 2004. Most of the fluctuations in the November indices are typical of those observed during earlier months and can be considered normal. However, over the interval December 01–03, an anomalous nonstationary process began to appear in the data for the SEMF~CO2 and it dropped down until it was back comparatively to the previous data over the interval December 04–15. Then the main nonstationary process began and it abruptly dropped down around December 16. It was not until the jump downward that the tsunamigenic earthquake with magnitude M9 in Indonesia (December 26, 2004) occurred, after about 11 days.

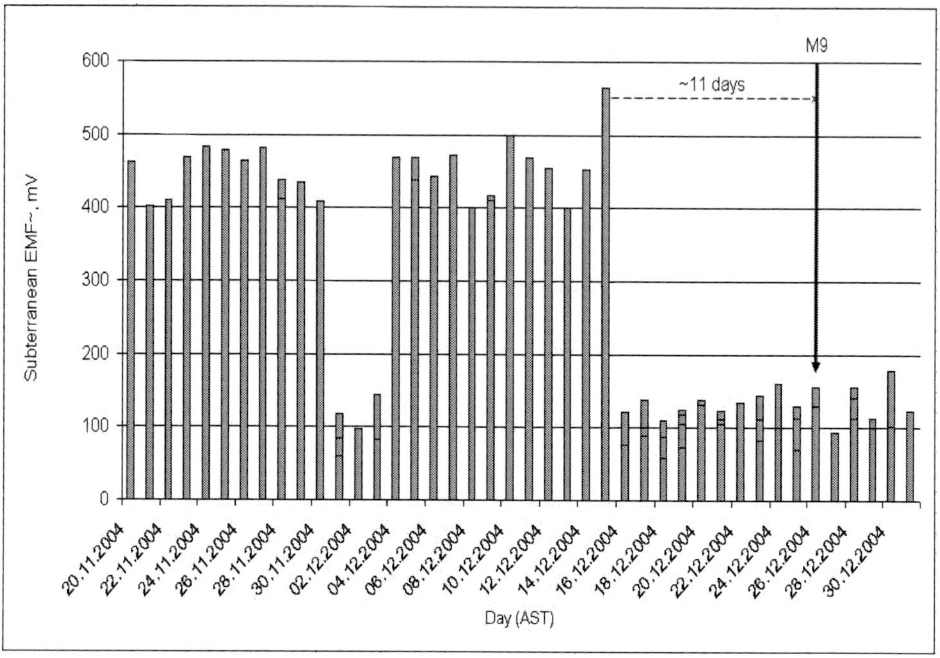

Figure 4. Column diagram of the largest daily signals of SEMF~ being taken in CO2~ channel over the interval November 20 to December 31, 2004. AST – apparent solar time in Petropavlovsk-Kamchatskiy, EMF~ – electromotive force of alternating currents, M9 – Indonesian earthquake magnitude (December 26, 2004), ~11 days – days interval of PE-signals abrupt jump downward before earthquake M9.

Figure 5 is a column diagram of the largest daily signals of SEMF~ being taken in CO3~ channel over the interval November 20 to December 31, 2004. Most of the fluctuations in the November indices are typical of those observed during earlier months and can be considered normal. However, over the interval December 1–3, an anomalous nonstationary process began to appear in the data for the SEMF~CO3 and it dropped down until it was back comparatively to the previous data over the interval December 4–15. Then the main nonstationary process began and it abruptly dropped down around December 16. It was not until the jump downward that the tsunamigenic earthquake with magnitude M9 in Indonesia (December 26, 2004) occurred, after about 11 days.

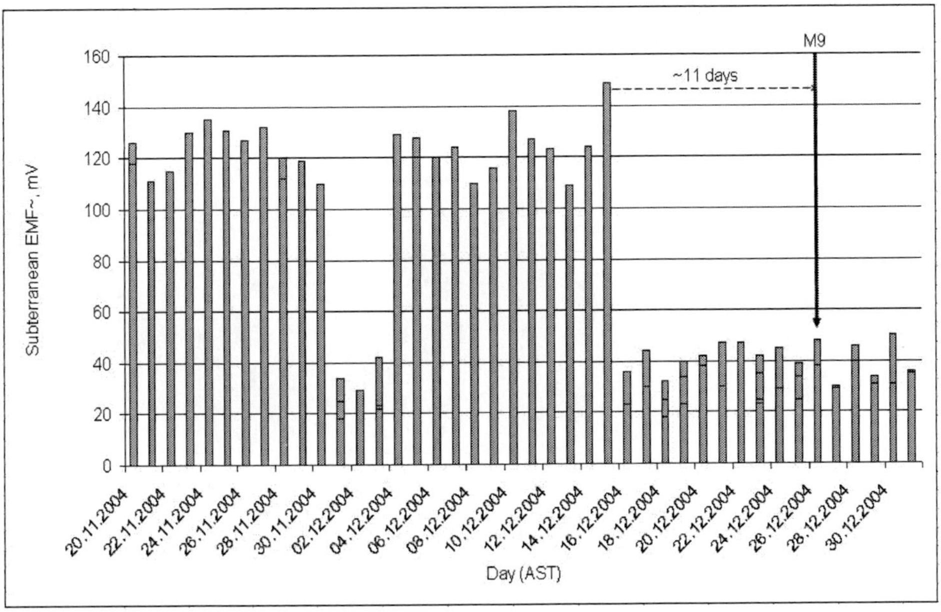

Figure 5. Column diagram of the largest daily signals of SEMF~ being taken in CO3~ channel over the interval November 20 to December 31, 2004. AST – apparent solar time in Petropavlovsk-Kamchatskiy, EMF~ – electromotive force of alternating currents, M9 – Indonesian earthquake magnitude (December 26, 2004), ~11 days – days interval of PE sign
als abrupt jump downward before earthquake M9.

Figure 6 is a column diagram of the largest daily signals of SEMF~ being taken in CO4~ channel over the interval November 20 to December 31, 2004. Most of the fluctuations in the November indices are typical of those observed during earlier months and can be considered normal. However, over the interval December 1–3, an anomalous nonstationary process began to appear in the data for the SEMF~CO4 and it dropped down until it was back comparatively to the previous data over the interval December 4–15. Then the main nonstationary process began and it abruptly dropped down around December 16. It was not until the jump downward that the tsunamigenic earthquake with magnitude M9 in Indonesia (December 26, 2004) occurred, after about 11 days.

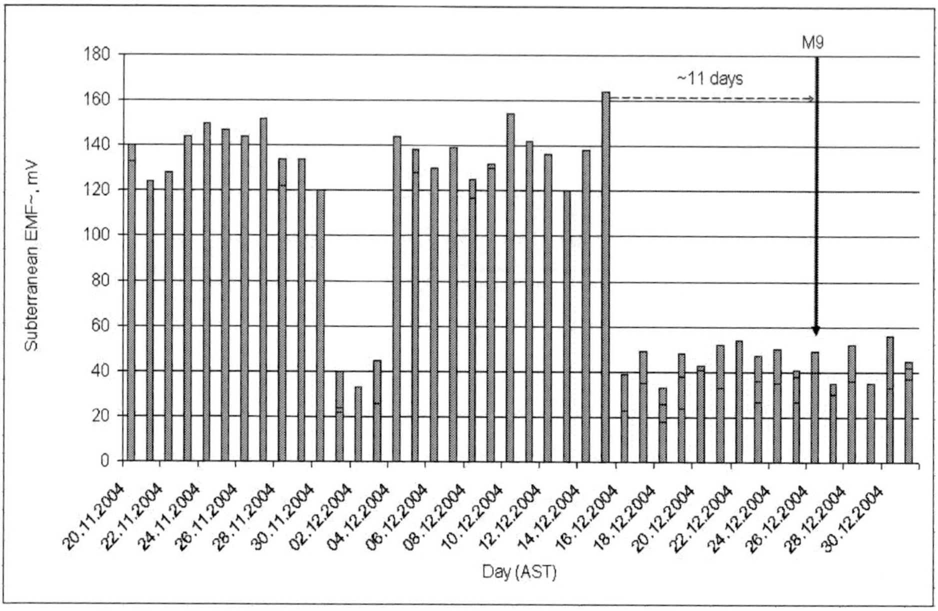

Figure 6. Column diagram of the largest daily signals of SEMF~ being taken in CO4~ channel over the interval November 20 to December 31, 2004. AST – apparent solar time in Petropavlovsk-Kamchatskiy, EMF~ – electromotive force of alternating currents, M9 – Indonesian earthquake magnitude (December 26, 2004), ~11 days – days interval of PE signals abrupt jump downward before earthquake M9.

Figure 7 is a column diagram of the largest daily signals of SEMF~ being taken in CO5~ channel over the interval November 20 to December 31, 2004. Most of the fluctuations in the November indices are typical of those observed during earlier months and can be considered normal. However, over the interval December 1–3, an anomalous nonstationary process began to appear in the data for the SEMF~CO5 and it dropped down until it was back comparatively to the previous data over the interval December 4–15. Then the main nonstationary process began and it abruptly dropped down around December 16. It was not until the jump downward that the tsunamigenic earthquake with magnitude M9 in Indonesia (December 26, 2004) occurred, after about 11 days.

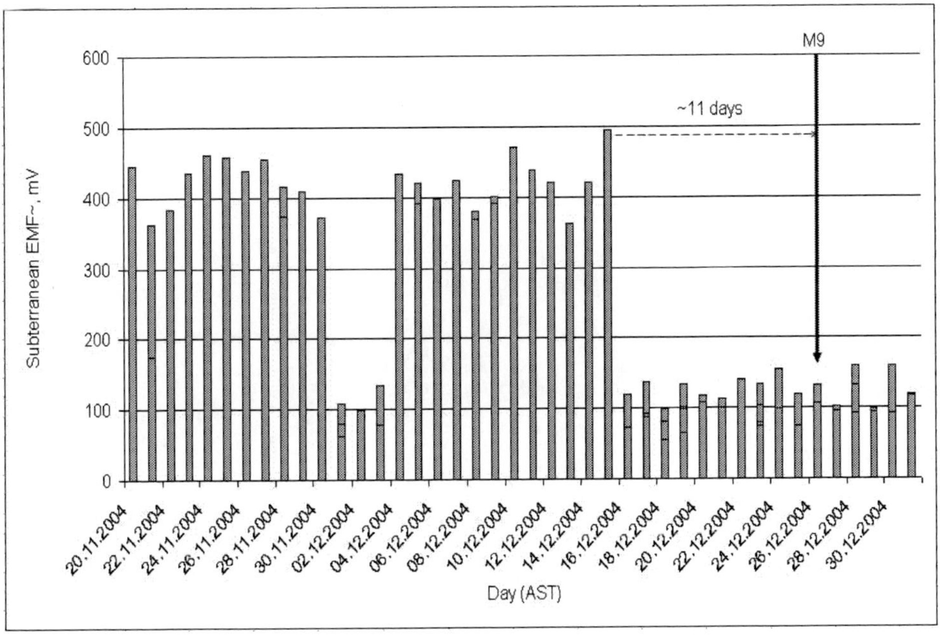

Figure 7. Column diagram of the largest daily signals of SEMF~ being taken in CO5~ channel over the interval November 20 to December 31, 2004. AST – apparent solar time in Petropavlovsk-Kamchatskiy, EMF~ – electromotive force of alternating currents, M9 – Indonesian earthquake magnitude (December 26, 2004), ~11 days – days interval of PE signals abrupt jump downward before earthquake M9.

Figure 8 is a column diagram of the largest daily signals of SEMF~ being taken in CO6~ channel over the interval November 20 to December 31, 2004. Most of the fluctuations in the November indices are typical of those observed during earlier months and can be considered normal. However, over the interval December 1–3, an anomalous nonstationary process began to appear in the data for the SEMF~CO6 and it dropped down until it was back comparatively to the previous data over the interval December 4–15. Then the main nonstationary process began and it abruptly dropped down around December 16. It was not until the jump downward that the tsunamigenic earthquake with magnitude M9 in Indonesia (December 26, 2004) occurred, after about 11 days.

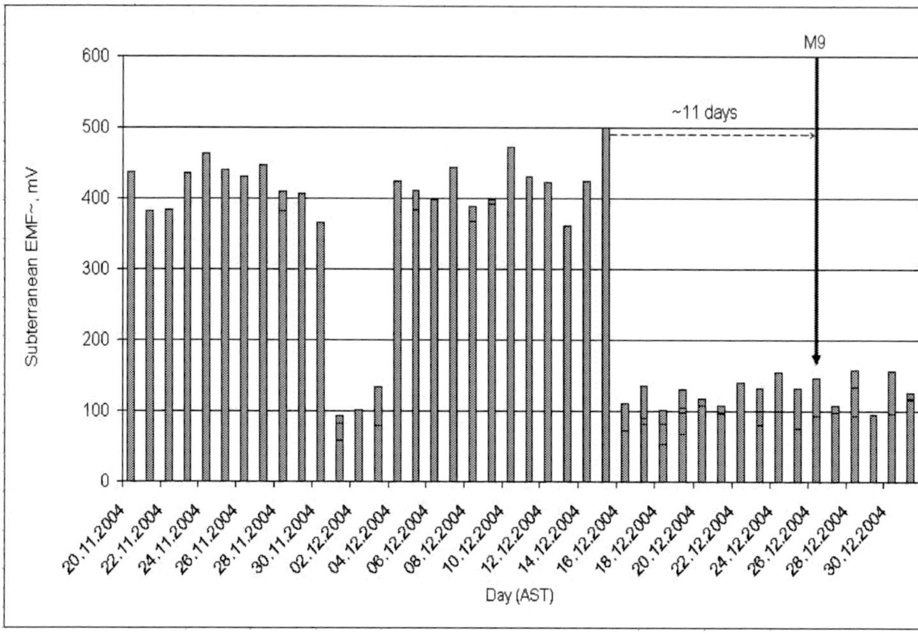

Figure 8. Column diagram of the largest daily signals of SEMF~ being taken in CO6~ channel over the interval November 20 to December 31, 2004. AST – apparent solar time in Petropavlovsk-Kamchatskiy, EMF~ – electromotive force of alternating currents, M9 – Indonesian earthquake magnitude (December 26, 2004), ~11 days – days interval of PE signals abrupt jump downward before earthquake M9.

Figure 9 is a column diagram of the largest daily signals of SEMF~ being taken in NE2C1~ channel over the interval November 20 to December 31, 2004. Most of the fluctuations in the November indices are typical of those observed during earlier months and can be considered normal. However, over the interval December 1–3, an anomalous nonstationary process began to appear in the data for the SEMF~ NE2C1 and it dropped down until it was back comparatively to the previous data over the interval December 4–15. Then the main nonstationary process began and it abruptly dropped down around December 16. It was not until the jump downward that the tsunamigenic earthquake with magnitude M9 in Indonesia (December 26, 2004) occurred, after about 11 days.

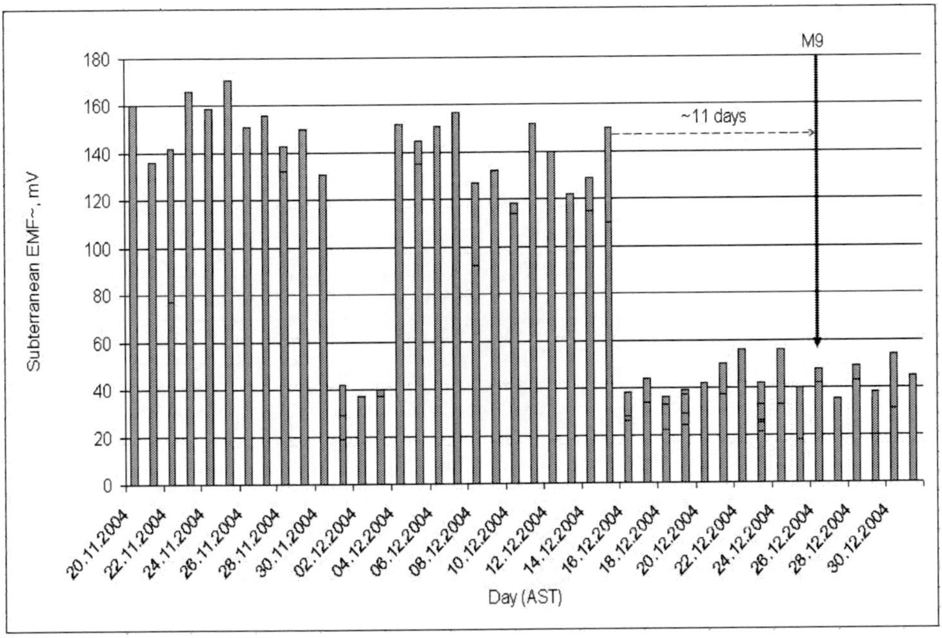

Figure 9. Column diagram of the largest daily signals of SEMF~ being taken in NE2C1~ channel over the interval November 20 to December 31, 2004. AST – apparent solar time in Petropavlovsk-Kamchatskiy, EMF~ – electromotive force of alternating currents, M9 – Indonesian earthquake magnitude (December 26, 2004), ~11 days – days interval of PE signals abrupt jump downward before earthquake M9.

Figure 10 is a column diagram of the largest daily signals of SEMF~ being taken in NE2C2~ channel over the interval November 20 to December 31, 2004. Most of the fluctuations in the November indices are typical of those observed during earlier months and can be considered normal. However, over the interval December 1–3, an anomalous nonstationary process began to appear in the data for the SEMF~ NE2C2 and it dropped down until it was back comparatively to the previous data over the interval December 4–15. Then the main nonstationary process began and it abruptly dropped down around December 16. It was not until the jump downward that the tsunamigenic earthquake with magnitude M9 in Indonesia (December 26, 2004) occurred, after about 11 days.

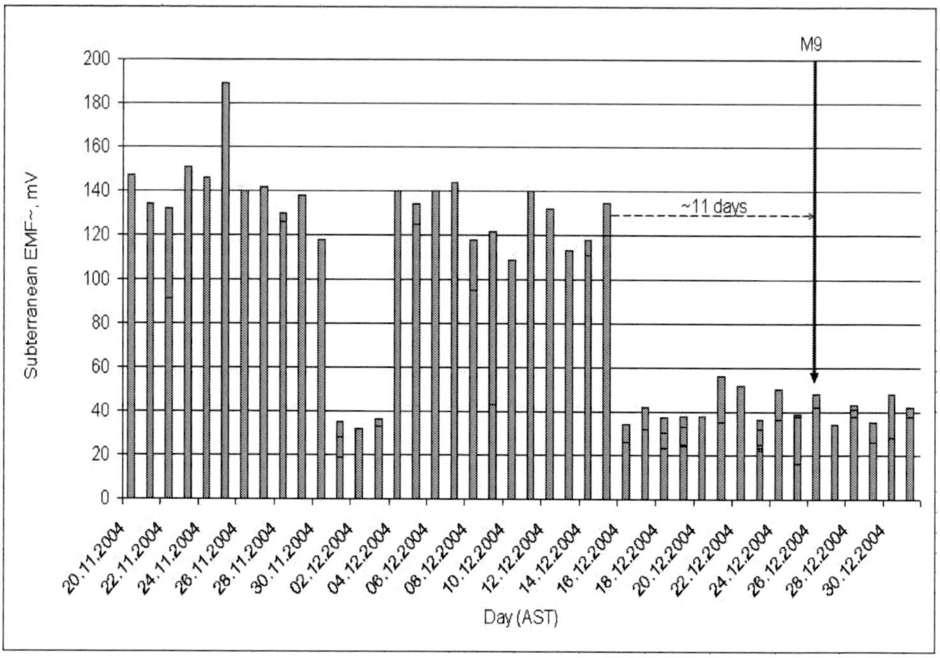

Figure 10. Column diagram of the largest daily signals of SEMF~ being taken in NE2C2~ channel over the interval November 20 to December 31, 2004. AST – apparent solar time in Petropavlovsk-Kamchatskiy, EMF~ – electromotive force of alternating currents, M9 – Indonesian earthquake magnitude (December 26, 2004), ~11 days – days interval of PE signals abrupt jump downward before earthquake M9.

Figure 11 is a column diagram of the largest daily signals of SEMF~ being taken in NE2C3~ channel over the interval November 20 to December 31, 2004. Most of the fluctuations in the November indices are typical of those observed during earlier months and can be considered normal. However, over the interval December 1–3, an anomalous nonstationary process began to appear in the data for the SEMF~ NE2C3 and it dropped down until it was back comparatively to the previous data over the interval December 4–15. Then the main nonstationary process began and it abruptly dropped down around December 16. It was not until the jump downward that the tsunamigenic earthquake with magnitude M9 in Indonesia (December 26, 2004) occurred, after about 11 days.

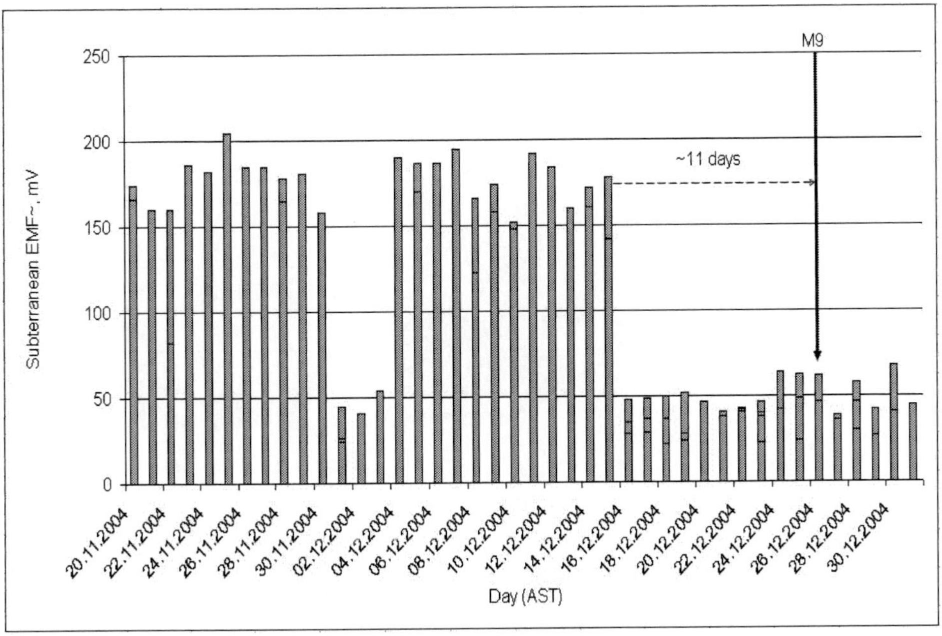

Figure 11. Column diagram of the largest daily signals of SEMF~ being taken in NE2C3~ channel over the interval November 20 to December 31, 2004. AST – apparent solar time in Petropavlovsk-Kamchatskiy, EMF~ – electromotive force of alternating currents, M9 – Indonesian earthquake magnitude (December 26, 2004), ~11 days – days interval of PE signals abrupt jump downward before earthquake M9.

Figure 12 is a column diagram of the largest daily signals of SEMF~ being taken in NE2C4~ channel over the interval November 20 to December 31, 2004. Most of the fluctuations in the November indices are typical of those observed during earlier months and can be considered normal. However, over the interval December 1–3, an anomalous nonstationary process began to appear in the data for the SEMF~ NE2C4 and it dropped down until it was back comparatively to the previous data over the interval December 4–15. Then the main nonstationary process began and it abruptly dropped down around December 16. It was not until the jump downward that the tsunamigenic earthquake with magnitude M9 in Indonesia (December 26, 2004) occurred, after about 11 days.

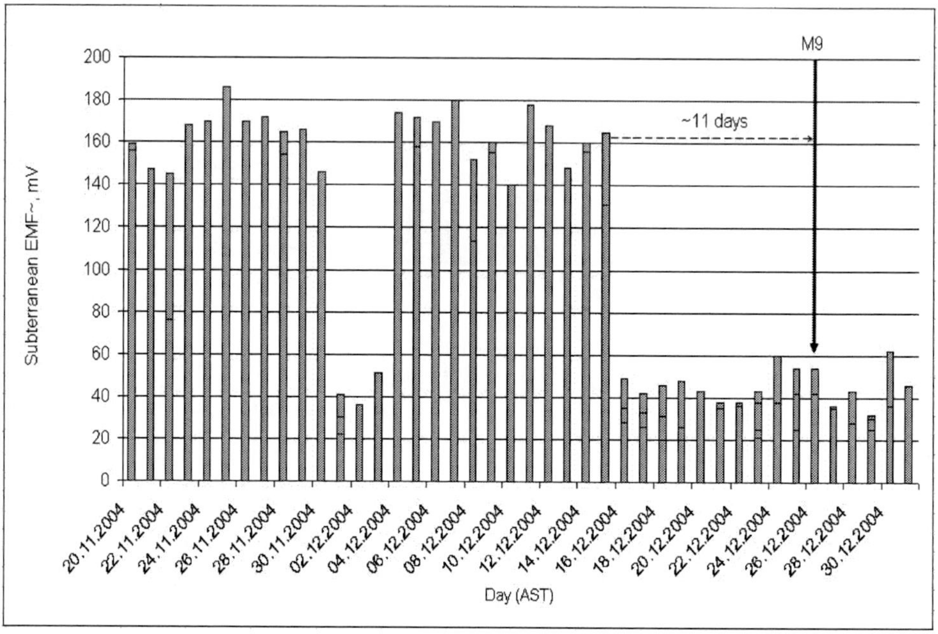

Figure 12. Column diagram of the largest daily signals of SEMF~ being taken in NE2C4~ channel over the interval November 20 to December 31, 2004. AST – apparent solar time in Petropavlovsk-Kamchatskiy, EMF~ – electromotive force of alternating currents, M9 – Indonesian earthquake magnitude (December 26, 2004), ~11 days – days interval of PE signals abrupt jump downward before earthquake M9.

Figure 13 is a column diagram of the largest daily signals of SEMF~ being taken in NE2C5~ channel over the interval November 20 to December 31, 2004. Most of the fluctuations in the November indices are typical of those observed during earlier months and can be considered normal. However, over the interval December 1–3, an anomalous nonstationary process began to appear in the data for the SEMF~ NE2C5 and it dropped down until it was back comparatively to the previous data over the interval December 4–15. Then the main nonstationary process began and it abruptly dropped down around December 16. It was not until the jump downward that the tsunamigenic earthquake with magnitude M9 in Indonesia (December 26, 2004) occurred, after about 11 days.

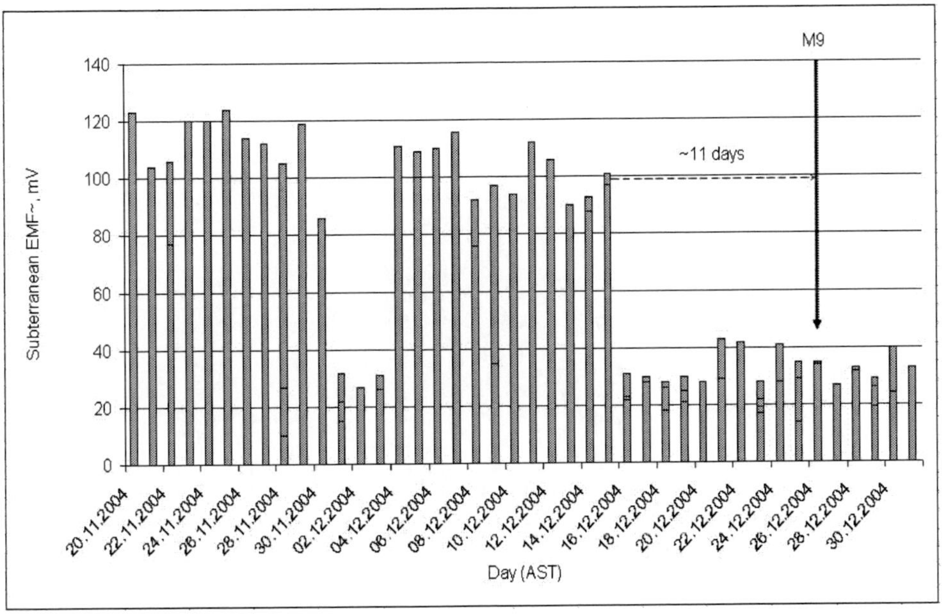

Figure 13. Column diagram of the largest daily signals of SEMF~ being taken in NE2C5~ channel over the interval November 20 to December 31, 2004. AST – apparent solar time in Petropavlovsk-Kamchatskiy, EMF~ – electromotive force of alternating currents, M9 – Indonesian earthquake magnitude (December 26, 2004), ~11 days – days interval of PE signals abrupt jump downward before earthquake M9.

Figure 14 is a column diagram of the largest daily signals of SEMF~ being taken in NE2C6~ channel over the interval November 20 to December 31, 2004. Most of the fluctuations in the November indices are typical of those observed during earlier months and can be considered normal. However, over the interval December 01–03 an anomalous nonstationary process began to appear in the data for the SEMF~ NE2C6 and it dropped down until it was back comparatively to the previous data over the interval December 4–15. Then the main nonstationary process began and it abruptly dropped down around December 16. It was not until the jump downward that the tsunamigenic earthquake with magnitude M9 in Indonesia (December 26, 2004) occurred, after about 11 days.

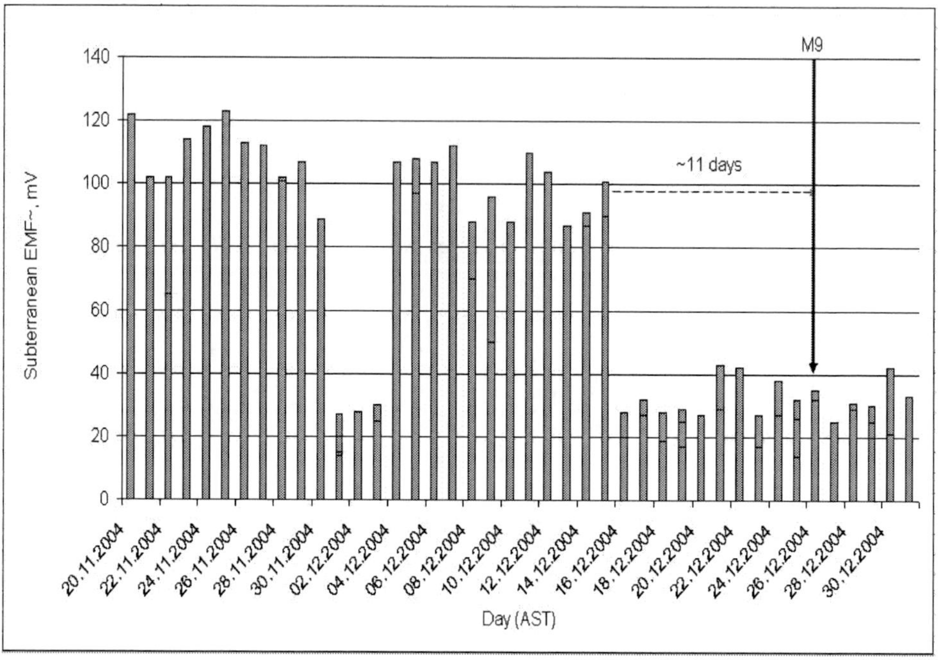

Figure 14. Column diagram of the largest daily signals of SEMF~ being taken in NE2C6~ channel over the interval November 20 to December 31, 2004. AST – apparent solar time in Petropavlovsk-Kamchatskiy, EMF~ – electromotive force of alternating currents, M9 – Indonesian earthquake magnitude (December 26, 2004), ~11 days – days interval of PE signals abrupt jump downward before earthquake M9.

Figure 15 is a column diagram of the largest daily signals of SEMF~ being taken in SW34~ channel over the interval November 20 to December 31, 2004. Most of the fluctuations in the November indices are typical of those observed during earlier months and can be considered normal. However, over the interval December 1–3, an anomalous nonstationary process began to appear in the data for the SEMF~ SW34 and it dropped down until it was back comparatively to the previous data over the interval December 4–15. Then the main nonstationary process began and it abruptly dropped down around December 16. It was not until the jump downward that the tsunamigenic earthquake with magnitude M9 in Indonesia (December 26, 2004) occurred, after about 11 days.

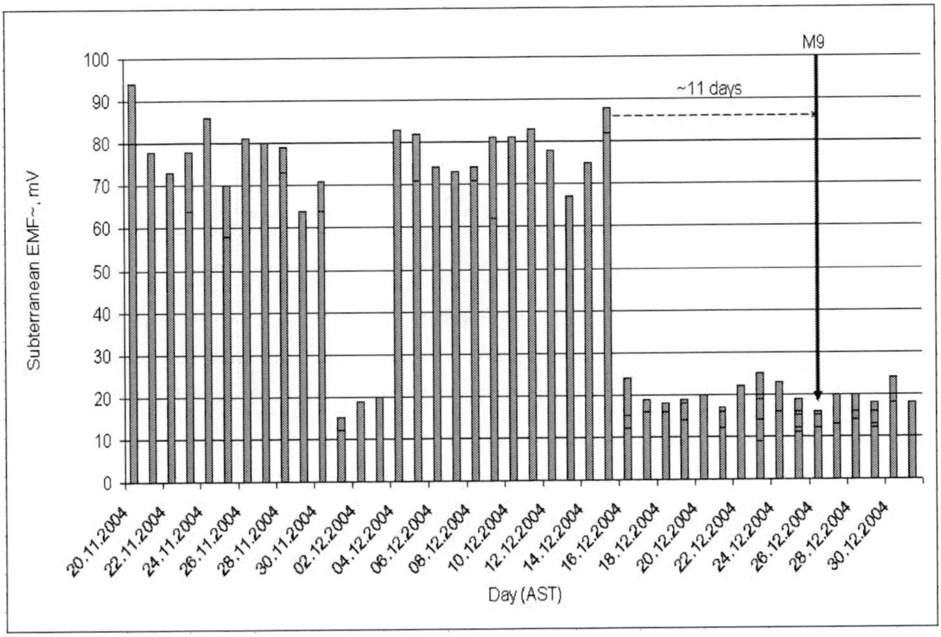

Figure 15. Column diagram of the largest daily signals of SEMF~ being taken in SW34~ channel over the interval November 20 to December 31, 2004. AST – apparent solar time in Petropavlovsk-Kamchatskiy, EMF~ – electromotive force of alternating currents, M9 – Indonesian earthquake magnitude (December 26, 2004), ~11 days – days interval of PE signals abrupt jump downward before earthquake M9.

Figure 16 is a column diagram of the largest daily signals of SEMF~ being taken in SW45~ channel over the interval November 20 to December 31, 2004. Most of the fluctuations in the November indices are typical of those observed during earlier months and can be considered normal. However, over the interval December 1–3, an anomalous nonstationary process began to appear in the data for the SEMF~SW45 and it dropped down until it was back main nonstationary process began and it abruptly dropped down around December 16. It was not until the jump downward that the tsunamigenic earthquake with magnitude M9 in Indonesia (December 26, 2004) occurred, after about 11 days.

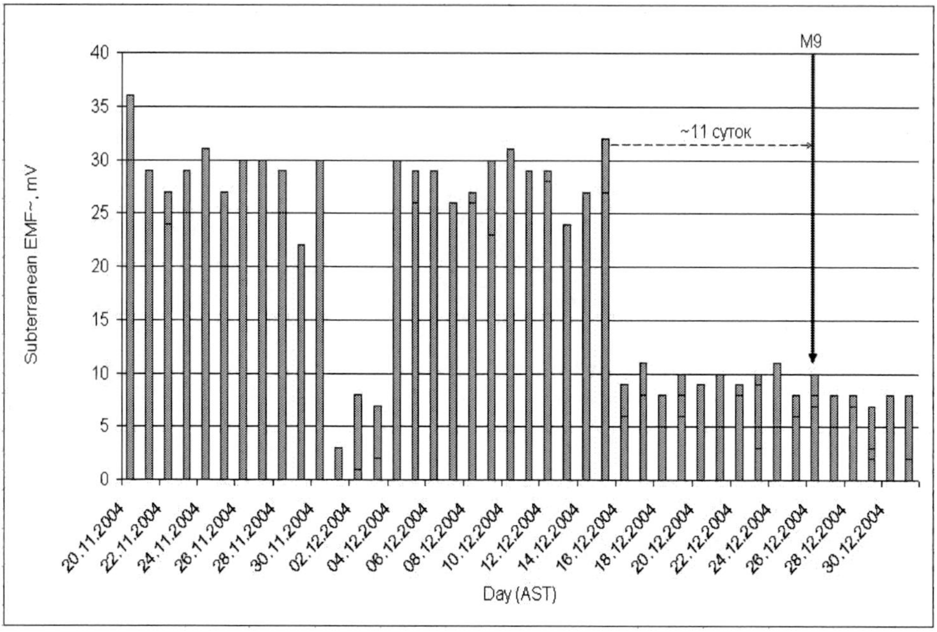

Figure 16. Column diagram of the largest daily signals of SEMF~ being taken in SW45~ channel over the interval November 20 to December 31, 2004. AST – apparent solar time in Petropavlovsk-Kamchatskiy, EMF~ – electromotive force of alternating currents, M9 – Indonesian earthquake magnitude (December 26, 2004), ~11 days – days interval of PE signals abrupt jump downward before earthquake M9.

Figure 17 shows a column diagram of the largest daily signals of SEMF~ being taken in SW56~ channel over the interval November 20 to December 31, 2004. Most of the fluctuations in the November indices are typical of those observed during earlier months and can be considered normal. However, over the interval December 1–3, an anomalous nonstationary process began to appear in the data for the SEMF~SW56 and it dropped down until it was back comparatively to the previous data over the interval December 4–15. Then the main nonstationary process began and it abruptly dropped down around December 16. It was not until the jump downward that the tsunamigenic earthquake with magnitude M9 in Indonesia (December 26, 2004) occurred, after about 11 days.

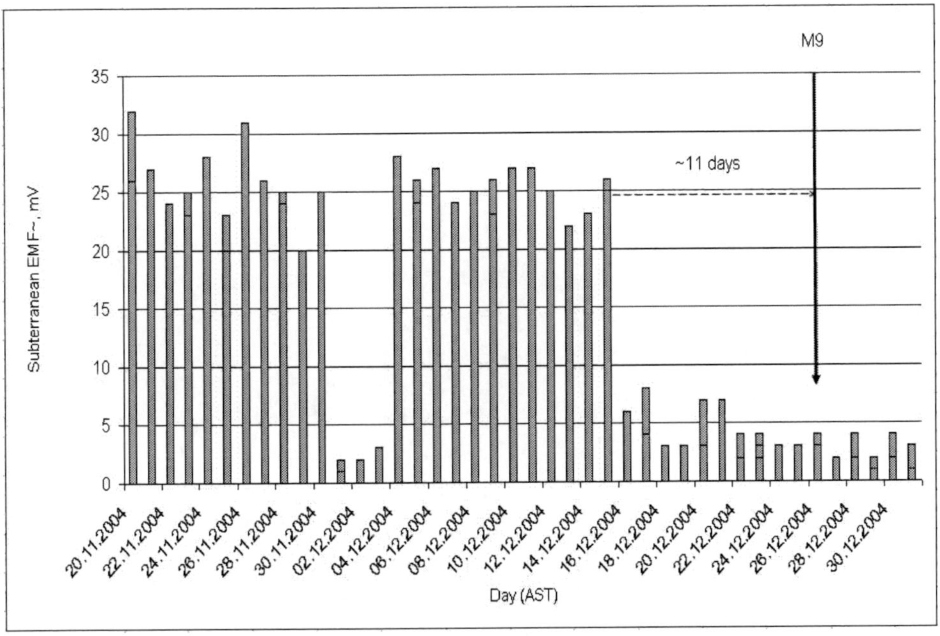

Figure 17. Column diagram of the largest daily signals of SEMF~ being taken in SW56~ channel over the interval November 20 to December 31, 2004. AST – apparent solar time in Petropavlovsk-Kamchatskiy, EMF~ – electromotive force of alternating currents, M9 – Indonesian earthquake magnitude (December 26, 2004), ~11 days – days interval of PE signals abrupt jump downward before earthquake M9.

Figure 18 shows a column diagram of the largest daily signals of SEMF~ being taken in SWO1~ channel over the interval November 20 to December 31, 2004. Most of the fluctuations in the November indices are typical of those observed during earlier months and can be considered normal. However, over the interval December 1–3, an anomalous nonstationary process began to appear in the data for the SEMF~SWO1 and it dropped down until it was back comparatively to the previous data over the interval December 4–15. Then the main nonstationary process began and it abruptly dropped down around December 16. It was not until the jump downward that the tsunamigenic earthquake with magnitude M9 in Indonesia (December 26, 2004) occurred, after about 11 days.

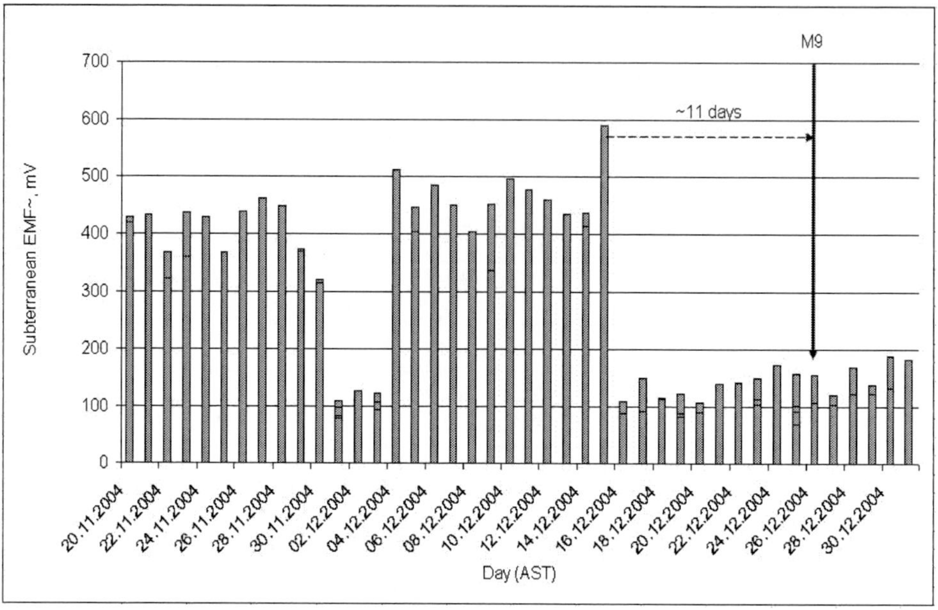

Figure 18. Column diagram of the largest daily signals of SEMF~ being taken in SWO1~ channel over the interval November 20 to December 31, 2004. AST – apparent solar time in Petropavlovsk-Kamchatskiy, EMF~ – electromotive force of alternating currents, M9 – Indonesian earthquake magnitude (December 26, 2004), ~11 days – days interval of PE signals abrupt jump downward before earthquake M9.

Figure 19 is a column diagram of the largest daily signals of SEMF~ being taken in SWO2~ channel over the interval November 20 to December 31, 2004. Most of the fluctuations in the November indices are typical of those observed during earlier months and can be considered normal. However, over the interval December 1–3, an anomalous nonstationary process began to appear in the data for the SEMF~SWO2 and it dropped down until it was back comparatively to the previous data over the interval December 4–15. Then the main nonstationary process began and it abruptly dropped down around December 16. It was not until the jump downward that the tsunamigenic earthquake with magnitude M9 in Indonesia (December 26, 2004) occurred, after about 11 days.

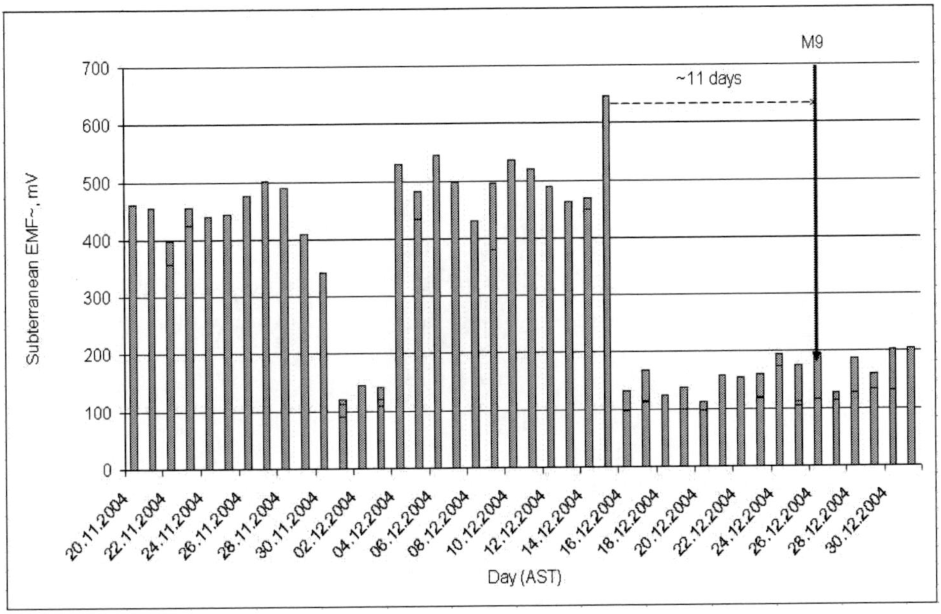

Figure 19. Column diagram of the largest daily signals of SEMF~ being taken in SWO2~ channel over the interval November 20 to December 31, 2004. AST – apparent solar time in Petropavlovsk-Kamchatskiy, EMF~ – electromotive force of alternating currents, M9 – Indonesian earthquake magnitude (December 26, 2004), ~11 days – days interval of PE signals abrupt jump downward before earthquake M9.

Figure 20 is a column diagram of the largest daily signals of SEMF~ being taken in SWO3~ channel over the interval November 20 to December 31, 2004. Most of the fluctuations in the November indices are typical of those observed during earlier months and can be considered normal. However, over the interval December 1–3, an anomalous nonstationary process began to appear in the data for the SEMF~SWO3 and it dropped down until it was back comparatively to the previous data over the interval December 4–15. Then the main nonstationary process began and it abruptly dropped down around December 16. It was not until the jump downward that the tsunamigenic earthquake with magnitude M9 in Indonesia (December 26, 2004) occurred, after about 11 days.

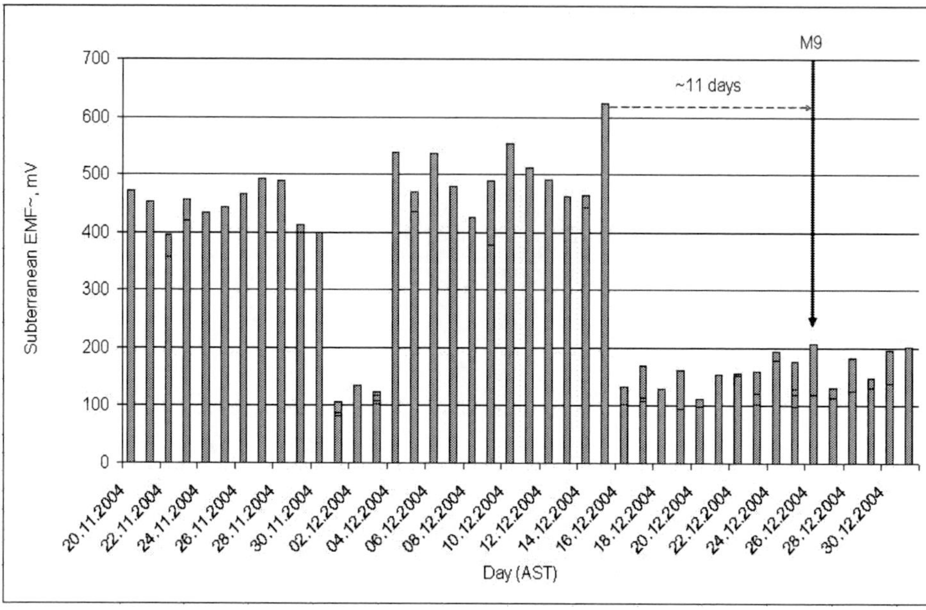

Figure 20. Column diagram of the largest daily signals of SEMF~ being taken in SWO3~ channel over the interval November 20 to December 31, 2004. AST – apparent solar time in Petropavlovsk-Kamchatskiy, EMF~ – electromotive force of alternating currents, M9 – Indonesian earthquake magnitude (December 26, 2004), ~11 days – days interval of PE signals abrupt jump downward before earthquake M9.

Figure 21 is a column diagram of the largest daily signals of SEMF~ being taken in SWO4~ channel over the interval November 20 to December 31, 2004. Most of the fluctuations in the November indices are typical of those observed during earlier months and can be considered normal. However, over the interval December 1–3, an anomalous nonstationary process began to appear in the data for the SEMF~SWO4 and it dropped down until it was back comparatively to the previous data over the interval December 4–15. Then the main nonstationary process began and it abruptly dropped down around December 16. It was not until the jump downward that the tsunamigenic earthquake with magnitude M9 in Indonesia (December 26, 2004) occurred, after about 11 days.

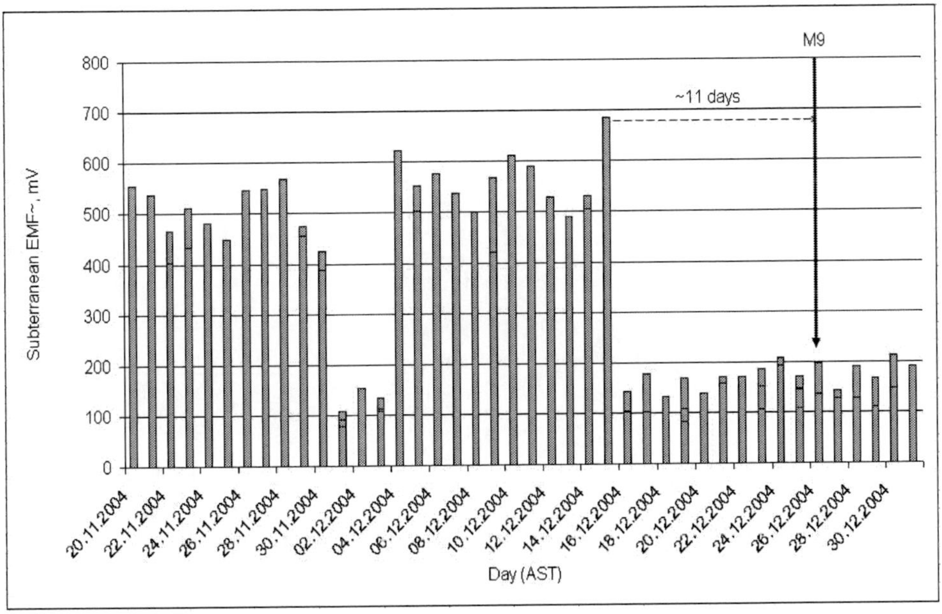

Figure 21. Column diagram of the largest daily signals of SEMF~ being taken in SWO4~ channel over the interval November 20 to December 31, 2004. AST – apparent solar time in Petropavlovsk-Kamchatskiy, EMF~ – electromotive force of alternating currents, M9 – Indonesian earthquake magnitude (December 26, 2004), ~11 days – days interval of PE signals abrupt jump downward before earthquake M9.

Figure 22 is a column diagram of the largest daily signals of SEMF~ being taken in SWO5~ channel over the interval November 20 to December 31, 2004. Most of the fluctuations in the November indices are typical of those observed during earlier months and can be considered normal. However, over the interval December 1–3, an anomalous nonstationary process began to appear in the data for the SEMF~SWO5 and it dropped down until it was back comparatively to the previous data over the interval December 4–15. Then the main nonstationary process began and it abruptly dropped down around December 16. It was not until the jump downward that the tsunamigenic earthquake with magnitude M9 in Indonesia (December 26, 2004) occurred, after about 11 days.

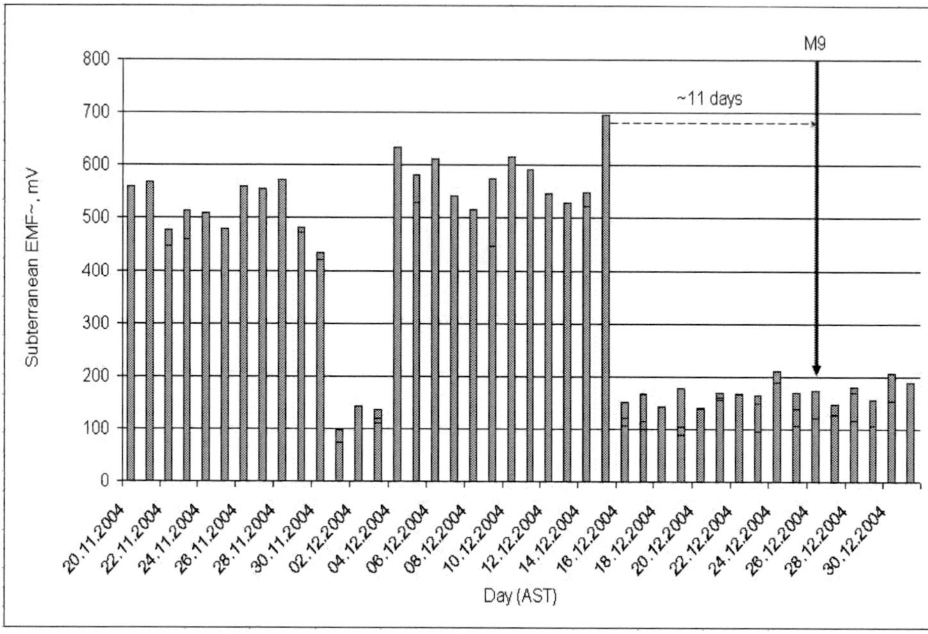

Figure 22. Column diagram of the largest daily signals of SEMF~ being taken in SWO5~ channel over the interval November 20 to December 31, 2004. AST – apparent solar time in Petropavlovsk-Kamchatskiy, EMF~ – electromotive force of alternating currents, M9 – Indonesian earthquake magnitude (December 26, 2004), ~11 days – days interval of PE signals abrupt jump downward before earthquake M9.

Figure 23 is a column diagram of the largest daily signals of SEMF~ being taken in SWO6~ channel over the interval November 20 to December 31, 2004. Most of the fluctuations in the November indices are typical of those observed during earlier months and can be considered normal. However, over the interval December 1–3, an anomalous nonstationary process began to appear in the data for the SEMF~SWO6 and it dropped down until it was back comparatively to the previous data over the interval December 4–15. Then the main nonstationary process began and it abruptly dropped down around December 16. It was not until the jump downward that the tsunamigenic earthquake with magnitude M9 in Indonesia (December 26, 2004) occurred, after about 11 days.

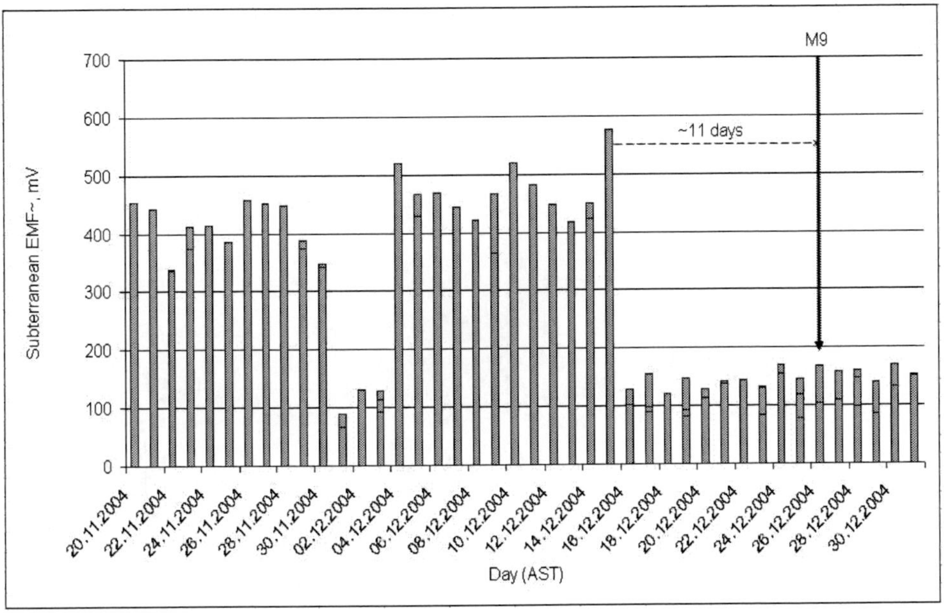

Figure 23. Column diagram of the largest daily signals of SEMF~ being taken in SWO6~ channel over the interval November 20 to December 31, 2004. AST – apparent solar time in Petropavlovsk-Kamchatskiy, EMF~ – electromotive force of alternating currents, M9 – Indonesian earthquake magnitude (December 26, 2004), ~11 days – days interval of PE signals abrupt jump downward before earthquake M9.

Figure 24 is a column diagram of the largest daily signals of SEMF~ being taken in NE~ channel over the interval November 20 to December 31, 2004. Most of the fluctuations in the November indices are typical of those observed during earlier months and can be considered normal. However, over the interval December 1–3 an anomalous nonstationary process began to appear in the data for the SEMF~NE and it dropped down until it was back comparatively to the previous data over the interval December 4–15. Then the main nonstationary process began and it abruptly dropped down around December 16. It was not until the jump downward that the tsunamigenic earthquake with magnitude M9 in Indonesia (December 26, 2004) occurred, after about 11 days.

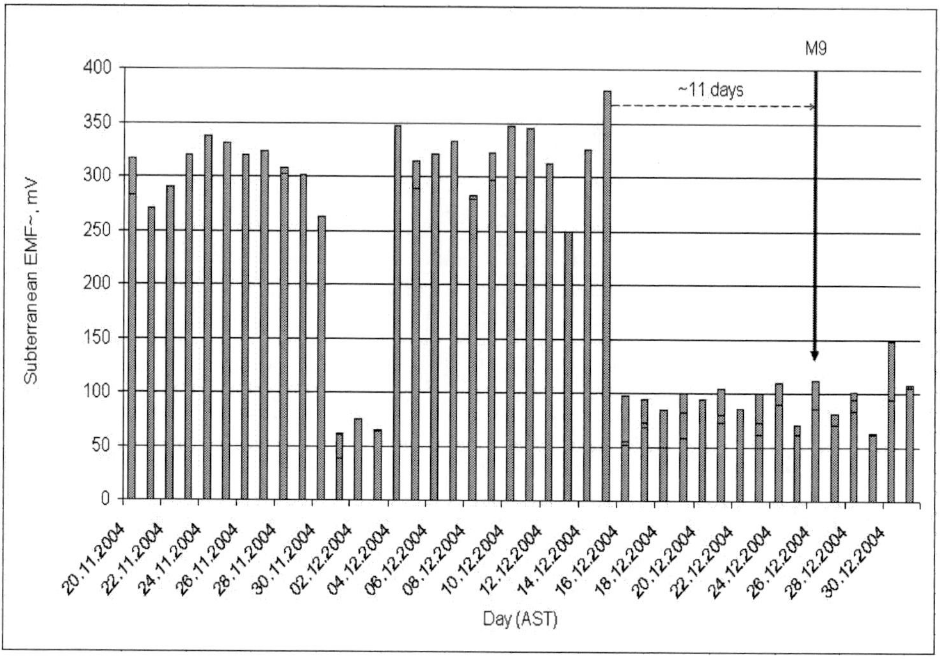

Figure 24. Column diagram of the largest daily signals of SEMF~ being taken in NE~ channel over the interval 2004 November 20 till 2004 December 31. AST – apparent solar time in Petropavlovsk-Kamchatskiy, EMF~ – electromotive force of alternating currents, M9 – Indonesian earthquake magnitude (December 26, 2004), ~11 days – days interval of PE signals abrupt jump downward before earthquake M9.

Moving to the catastrophic China earthquake (May 12, 2008, M7.9), Figure 25 illustrates a column diagram of the largest daily signals of SEMF~ taken in the NE~ channel over the interval May 28, 2004 to May 31, 2008. The column diagram shows large signals of SEMF~NE during more than six months directly before the tsunamigenic Indonesia earthquake (December 26, 2004, M9). The maximum value was 380 mV on 2004 December 15, i.e., directly before the rapid phase of earthquake inception in Indonesia. The next significant signal rising of SEMF~NE was observed in 2008; the maximum value was 347 mV on March 2, 2008, i.e., about two months directly before the rapid phase of earthquake inception in China (May 12, 2008, M7.9).

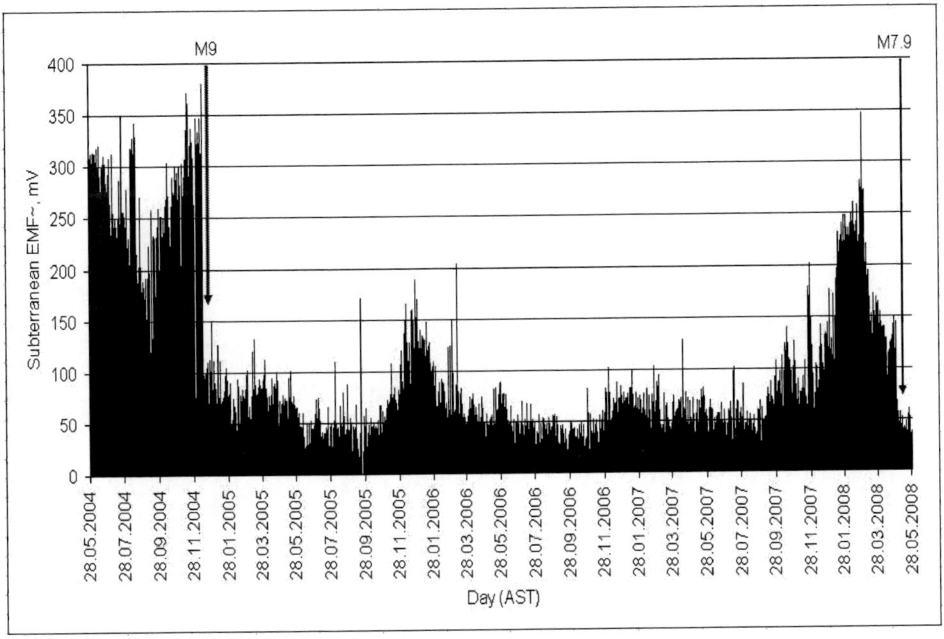

Figure 25. Column diagram of the largest daily signals of SEMF~ being taken in NE~ channel over the interval May 28, 2004 to May 31, 2008. AST – apparent solar time in Petropavlovsk-Kamchatskiy, EMF~ – electromotive force of alternating currents, M9 – Indonesian earthquake magnitude (December 26, 2004), M7.9 – Chinese earthquake magnitude (May 12, 2008).

To discuss the rapid onset of the China earthquake (May 12, 2008), Figure 26 shows a column diagram of the largest daily signals of SEMF~ being taken in the NE~ channel over the interval April 14 to May 20, 2008. Most of the fluctuations in the April indices are typical and can be considered normal. However, an anomalous main nonstationary process began to appear in the data for the SEMF~NE and it abruptly dropped down around May 1. It was not until the jump downward that the catastrophic earthquake with magnitude M7.9 in China (May 12, 2008) occurred, after about 12 days.

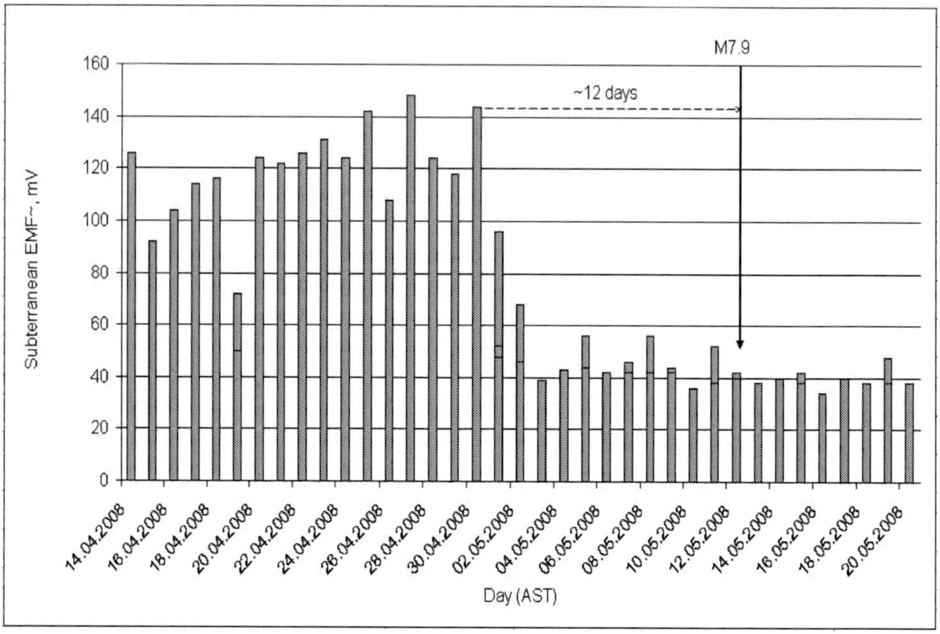

Figure 26. Column diagram of the largest daily signals of SEMF~ being taken in NE~ channel over the interval April 14 to May 20, 2008. AST – apparent solar time in Petropavlovsk-Kamchatskiy, EMF~ – electromotive force of alternating currents, M7.9 – Chinese earthquake magnitude (May 12, 2008), ~12 days – days interval of PE signals abrupt jump down before earthquake M7.9.

Moving to the catastrophic China earthquake (May 12, 2008, M7.9), Figure 27 illustrates a column diagram of the largest daily signals of SEMF~ being taken in C23~ channel over the interval May 28, 2004 to May 31, 2008. The column shows large signals of SEMF~C23 during more than six months directly before tsunamigenic Indonesia earthquake (December 26, 2004, M9). The maximum value was 384 mV on November 17, 2004, i.e., directly a month before speedy phase of earthquake preparation in Indonesia. The next significant signal rising of SEMF~C23 was observed in 2008; the maximum value was 236 mV on March 2, 2008, i.e., about two months directly before the rapid phase of earthquake inception in China (May 12, 2008, M7.9).

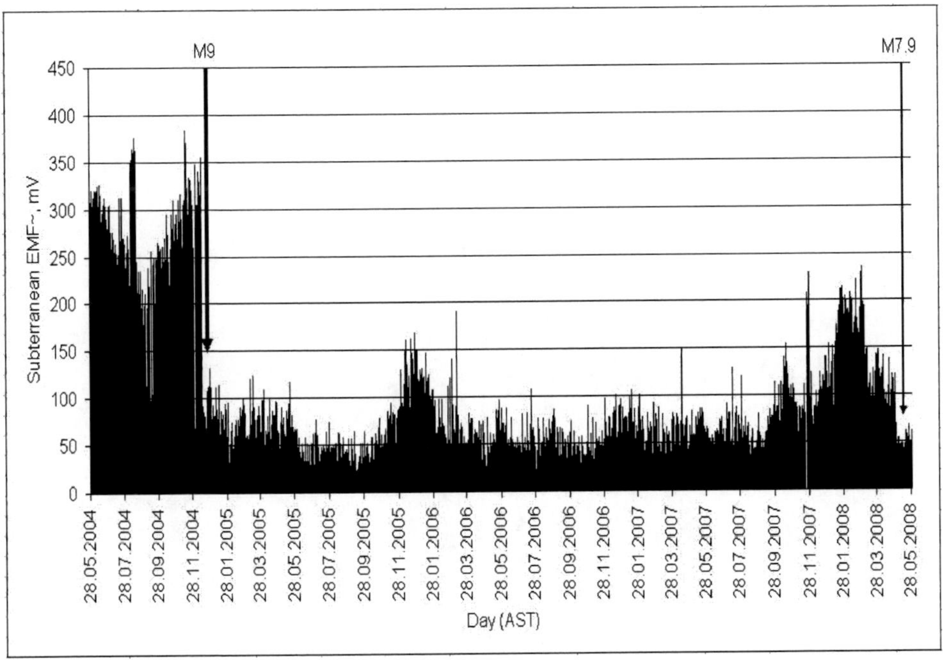

Figure 27. Column diagram of the largest daily signals of SEMF~ being taken in C23~ channel over the interval May 28, 2004 to May 31, 2008. AST – apparent solar time in Petropavlovsk-Kamchatskiy, EMF~ – electromotive force of alternating currents, M9 – Indonesian earthquake magnitude (December 26, 2004), M7.9 – Chinese earthquake magnitude (May 12, 2008).

To discuss the rapid onset of the China earthquake (May 12, 2008), Figure 28 shows a column diagram of the largest daily signals of SEMF~ being taken in the C23~ channel over the interval April 14 to May 20, 2008. Most of the fluctuations in the April indices are typical and can be considered normal. However, an anomalous main nonstationary process began to appear in the data for the SEMF~C23 and it abruptly dropped down around May 1. It was not until the jump downward that the catastrophic earthquake with magnitude M7.9 in China (May 12, 2008) occurred, after about 12 days.

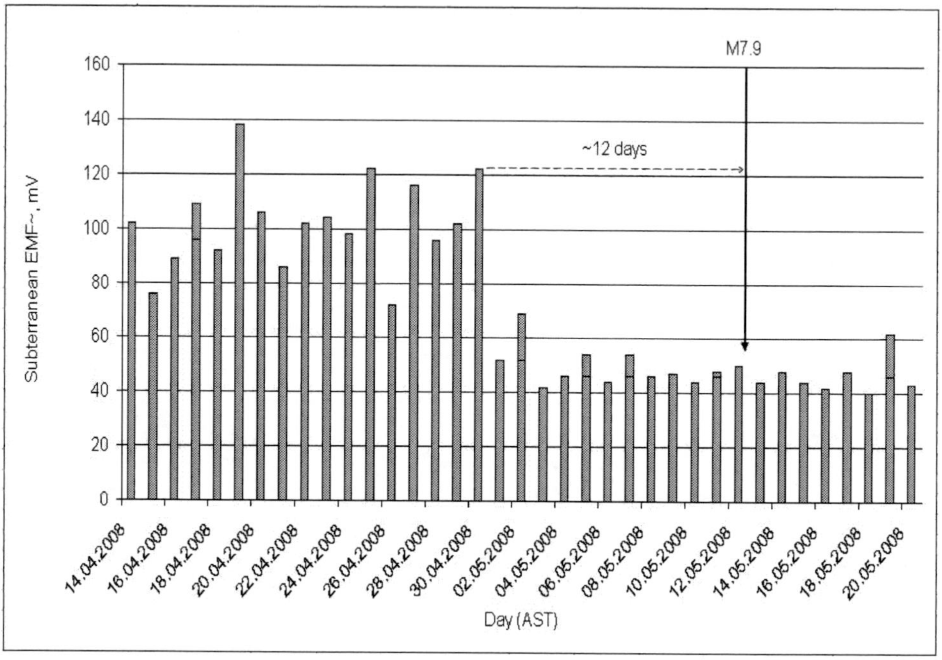

Figure 28. Column diagram of the largest daily signals of SEMF~ being taken in C23~ channel over the interval April 14 to May 20, 2008. AST – apparent solar time in Petropavlovsk-Kamchatskiy, EMF~ – electromotive force of alternating currents, M7.9 – Chinese earthquake magnitude (May 12, 2008), ~12 days – days interval of PE signals abrupt jump down before earthquake M7.9.

Moving to the catastrophic China earthquake (May 12, 2008, M7.9), Figure 29 is a column diagram of the largest daily signals of SEMF~ being taken in C45~ channel over the interval May 28, 2004 to May 31, 2008. The column diagram shows large signals of SEMF~C45 during more than sixs months directly before the tsunamigenic Indonesia earthquake (December 26, 2004, M9). The maximum value was 354 mV on December 15, 2004, i.e., directly before the rapid phase of earthquake inception in Indonesia. The next significant signal rising of SEMF~C45 was observed in 2008; the maximum value was 208 mV on March 2, 2008, i.e., about two months directly before the rapid phase of earthquake inception in China (May 12, 2008, M7.9).

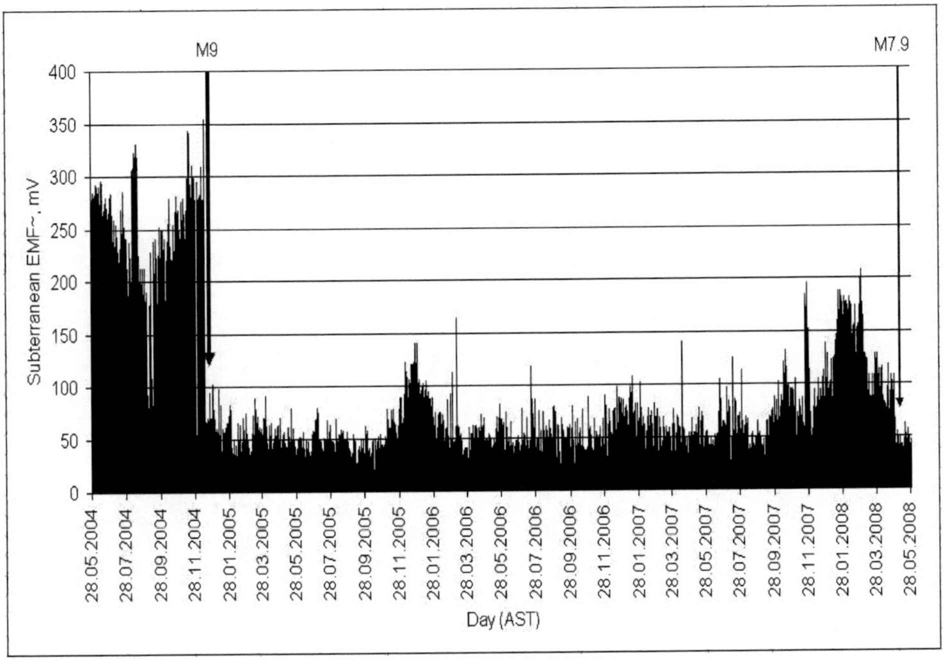

Figure 29. Column diagram of the largest daily signals of SEMF~ being taken in C45~ channel over the interval May 28, 2004 to May 31, 2008. AST – apparent solar time in Petropavlovsk-Kamchatskiy, EMF~ – electromotive force of alternating currents, M9 – Indonesian earthquake magnitude (December 26, 2004), M7.9 – Chinese earthquake magnitude (May 12, 2008).

To discuss the rapid onset of the China earthquake (May 12, 2008), Figure 30 shows a column diagram of the largest daily signals of SEMF~ taken in the C45~ channel over the interval April 14 to May 20, 2008. Most of the fluctuations in the April indices are typical and can be considered normal. However, an anomalous main nonstationary process began to appear in the data for the SEMF~C45 and it abruptly dropped down around May 1. It was not until the jump downward that the catastrophic earthquake with magnitude M7.9 in China (May 12, 2008) occurred, after about 12 days.

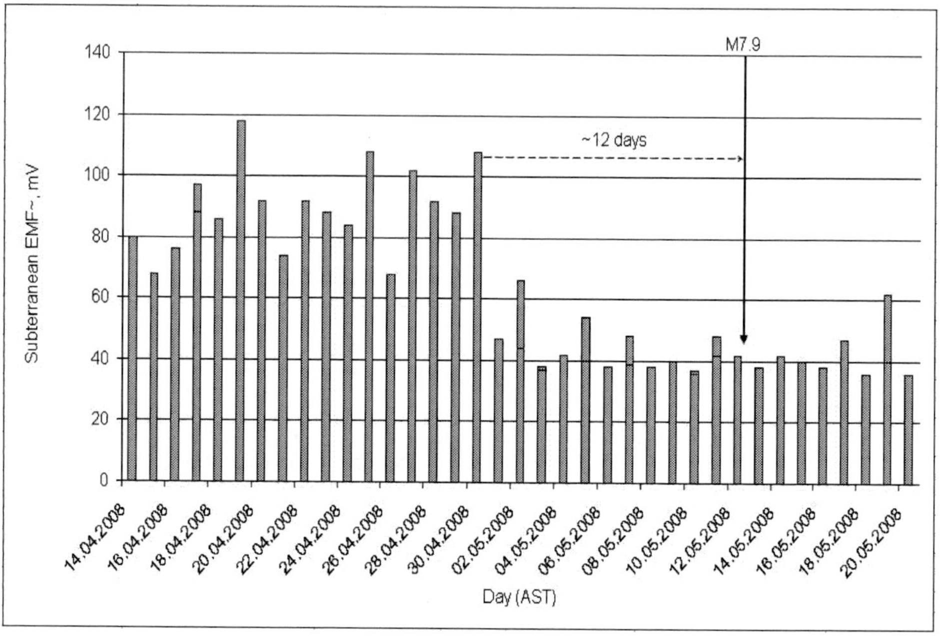

Figure 30. Column diagram of the largest daily signals of SEMF~ being taken in C45~ channel over the interval April 14 to May 20, 2008. AST – apparent solar time in Petropavlovsk-Kamchatskiy, EMF~ – electromotive force of alternating currents, M7.9 – Chinese earthquake magnitude (May 12, 2008), ~12 days – days interval of PE signals abrupt jump down before earthquake M7.9.

Moving to the catastrophic China earthquake (May 12, 2008, M7.9), Figure 31 illustrates a column diagram of the largest daily signals of SEMF~ taken in CO1~ channel over the interval May 28, 2004 to May 31, 2008. The column diagram shows large signals of SEMF~CO1 during more than six months directly before the tsunamigenic Indonesia earthquake (December 26, 2004, M9). The maximum value was 572 mV on December 15, 2004, i.e., directly before the rapid phase of earthquake inception in Indonesia. The next significant signal rising of SEMF~CO1 was observed in 2008; the maximum value was 337 mV on March 2, 2008, i.e., about two months directly before the rapid phase of earthquake inception in China (May 12, 2008, M7.9).

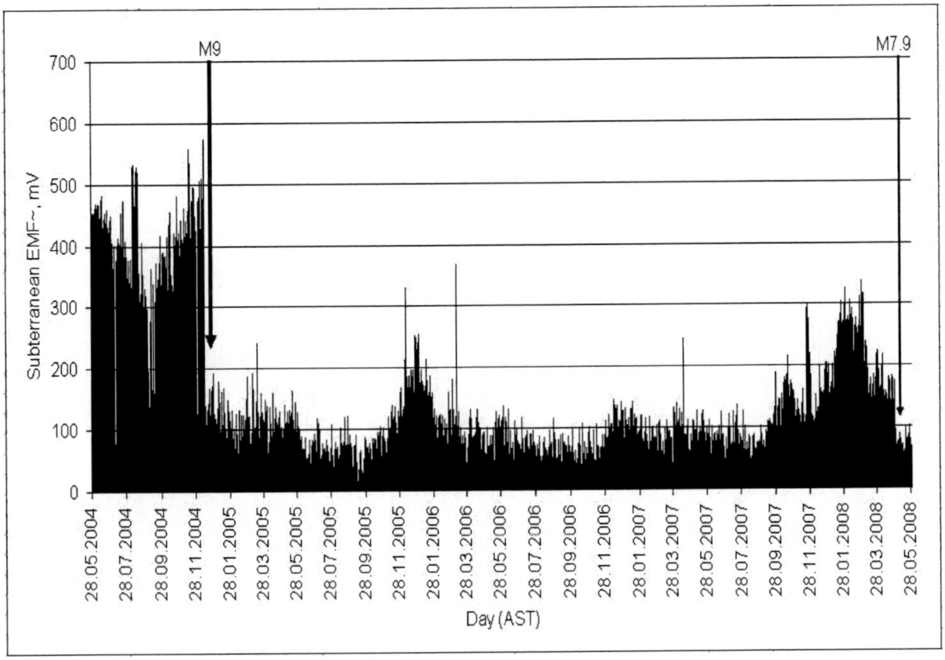

Figure 31. Column diagram of the largest daily signals of SEMF~ being taken in CO1~ channel over the interval May 28, 2004 to May 31, 2008. AST – apparent solar time in Petropavlovsk-Kamchatskiy, EMF~ – electromotive force of alternating currents, M9 – Indonesian earthquake magnitude (December 26, 2004), M7.9 – Chinese earthquake magnitude (May 12, 2008).

To discuss rapid onset of the China earthquake (May 12, 2008), Figure 32 shows a column diagram of the largest daily signals of SEMF~ taken in the CO1~ channel over the interval April 14 to May 20, 2008. Most of the fluctuations in the April indices are typical and can be considered normal. However, an anomalous main nonstationary process began to appear in the data for the SEMF~CO1 and it abruptly dropped down around May 1. It was not until the jump downward that the catastrophic earthquake with magnitude M7.9 in China (May 12, 2008) occurred, after about 12 days.

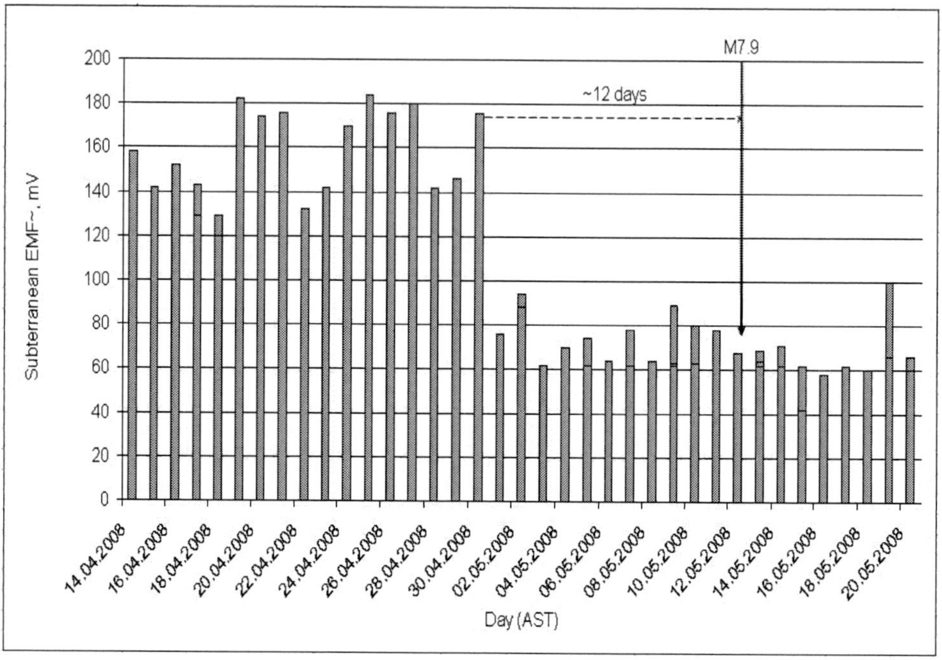

Figure 32. Column diagram of the largest daily signals of SEMF~ being taken in CO1~ channel over the interval April 14 to May 20, 2008. AST – apparent solar time in Petropavlovsk-Kamchatskiy, EMF~ – electromotive force of alternating currents, M7.9 – Chinese earthquake magnitude (May 12, 2008), ~12 days – days interval of PE signals abrupt jump down before earthquake M7.9.

Moving to the catastrophic China earthquake (May 12, 2008, M7.9), Figure 33 illustrates a column diagram of the largest daily signals of SEMF~ taken in the CO2~ channel over the interval May 28, 2004 to May 31, 2008. The column diagram shows large signals of SEMF~CO2 during more than six months directly before the tsunamigenic Indonesia earthquake (December 26, 2004, M9). The maximum value was 565 mV on December 15, 2004, i.e., directly before the rapid phase of earthquake inception in Indonesia. The next significant signal rising of SEMF~CO2 was observed in 2008; the maximum value was 316 mV on March 02, 2008, i.e., about two months directly before the rapid phase of earthquake inception in China (May 12, 2008, M7.9).

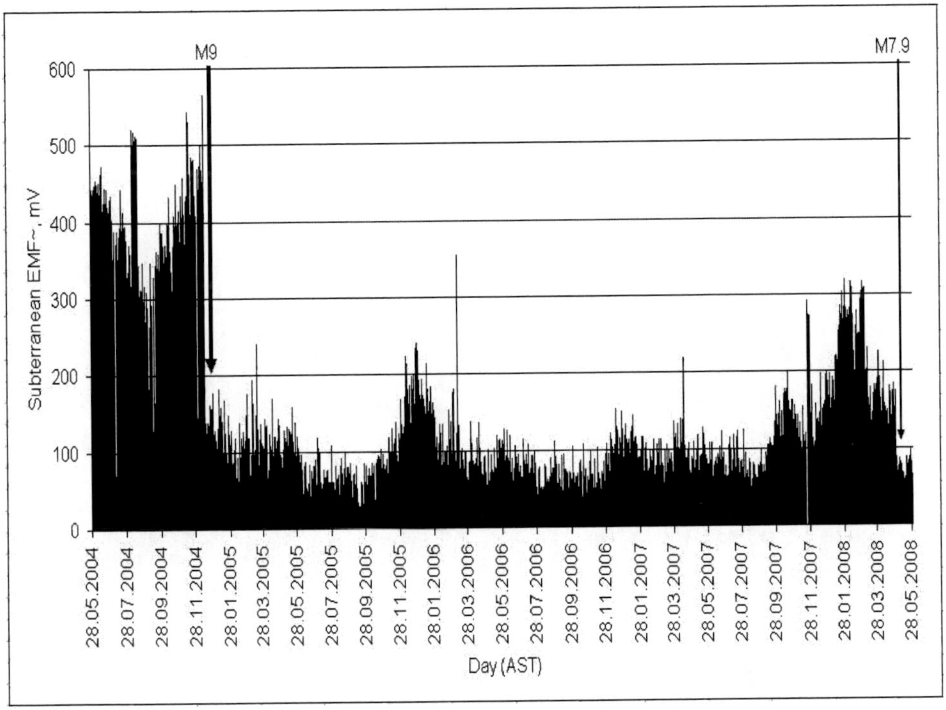

Figure 33. Column diagram of the largest daily signals of SEMF~ being taken in CO2~ channel over the interval May 28, 2004 to May 31, 2008. AST – apparent solar time in Petropavlovsk-Kamchatskiy, EMF~ – electromotive force of alternating currents, M9 – Indonesian earthquake magnitude (December 26, 2004), M7.9 – Chinese earthquake magnitude (May 12, 2008).

Figure 34 is a column diagram of the largest daily signals of SEMF~ taken in the CO2~ channel over the interval April 14 to May 20, 2008. Most of the fluctuations in the April indices are typical and can be considered normal. However, an anomalous main nonstationary process began to appear in the data for the SEMF~CO2 and it abruptly dropped down around May 1. It was not until the jump downward that the catastrophic earthquake with magnitude M7.9 in China (May 12, 2008) occurred, after about 12 days.

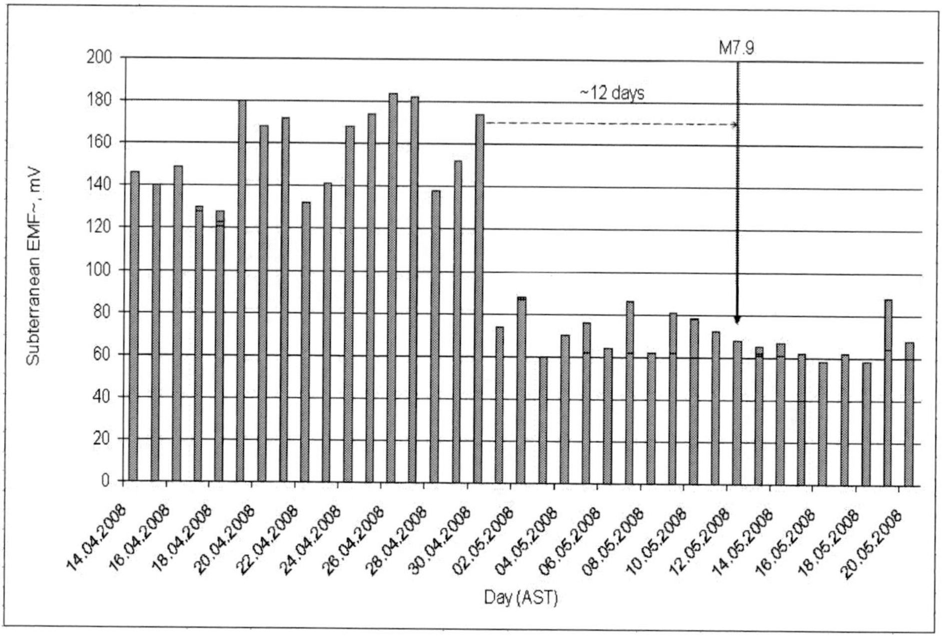

Figure 34. Column diagram of the largest daily signals of SEMF~ being taken in CO2~ channel over the interval April 14 to May 20, 2008. AST – apparent solar time in Petropavlovsk-Kamchatskiy, EMF~ – electromotive force of alternating currents, M7.9 – Chinese earthquake magnitude (May 12, 2008), ~12 days – days interval of PE signals abrupt jump down before earthquake M7.9.

Moving to the catastrophic China earthquake (May 12, 2008, M7.9), Figure 35 illustrates a column diagram of the largest daily signals of SEMF~ being taken in CO3~ channel over the interval May 28, 2004 to May 31, 2008. The column diagram shows large signals of SEMF~CO3 during more than six months directly before the tsunamigenic Indonesia earthquake (December 26, 2004, M9). The maximum value was 152 mV on November 17, 2004, i.e., about a month before the rapid phase of earthquake inception in Indonesia. The next significant signal rising of SEMF~CO3 was observed in 2008; the maximum value was 88 mV on February 10, 2008, i.e., about three months directly before the rapid phase of earthquake inception in China (May 12, 2008, M7.9).

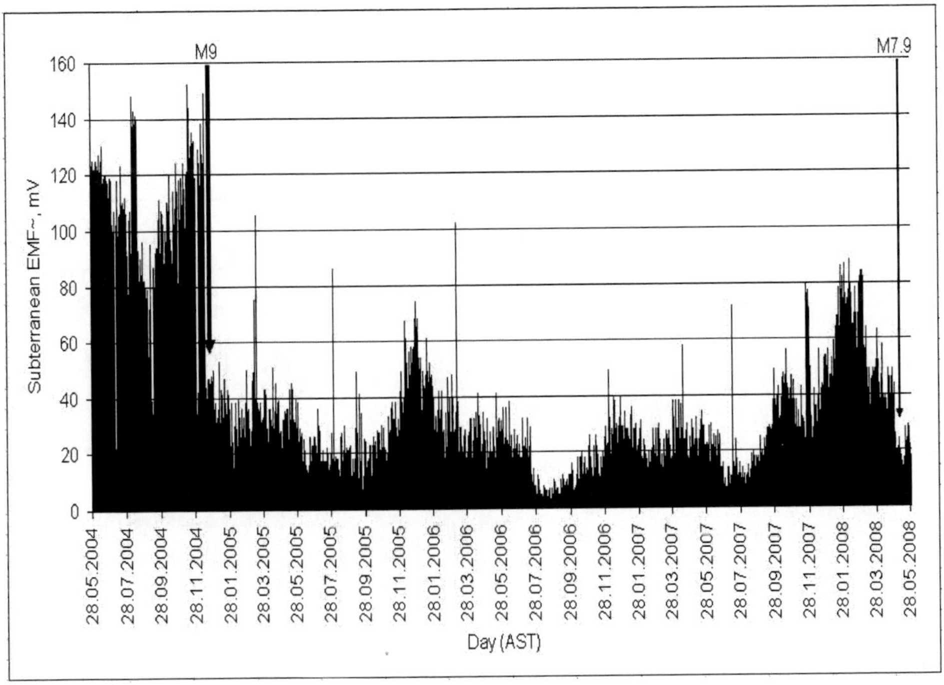

Figure 35. Column diagram of the largest daily signals of SEMF~ being taken in CO3~ channel over the interval May 28, 2004 to May 31, 2008. AST – apparent solar time in Petropavlovsk-Kamchatskiy, EMF~ – electromotive force of alternating currents, M9 – Indonesian earthquake magnitude (December 26, 2004), M7.9 – Chinese earthquake magnitude (May 12, 2008).

Figure 36 is a column diagram of the largest daily signals of SEMF~ being taken in CO3~ channel over the interval April 14 to May 20, 2008. Most of the fluctuations in the April indices are typical and can be considered normal. However, an anomalous main nonstationary process began to appear in the data for the SEMF~CO3 and it abruptly dropped down around May 1. It was not until the jump downward that the catastrophic earthquake with magnitude M7.9 in China (May 12, 2008) occurred, after about 12 days.

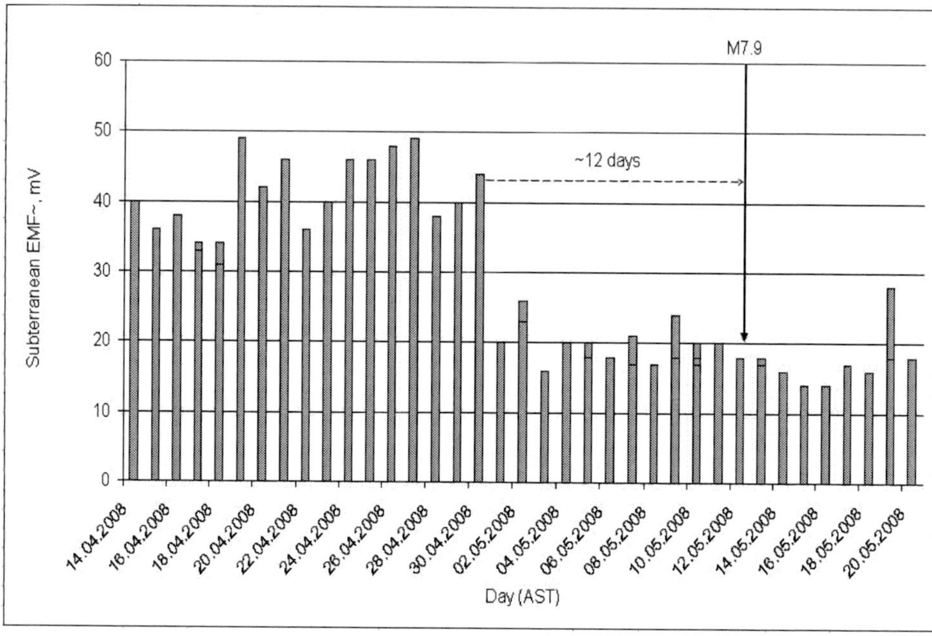

Figure 36. Column diagram of the largest daily signals of SEMF~ being taken in CO3~ channel over the interval April 14 to May 20, 2008. AST – apparent solar time in Petropavlovsk-Kamchatskiy, EMF~ – electromotive force of alternating currents, M7.9 – Chinese earthquake magnitude (May 12, 2008), ~12 days – days interval of PE signals abrupt jump down before earthquake M7.9.

Moving to the catastrophic China earthquake (May 12, 2008, M7.9), Figure 37 illustrates a column diagram of the largest daily signals of SEMF~ taken in CO4~ channel over the interval May 28, 2004 to May 31, 2008. The column diagram shows large signals of SEMF~CO4 during more than six months directly before the tsunamigenic Indonesia earthquake (December 26, 2004, M9). The maximum value was 170 mV on November 17, 2004, i.e., a month before the rapid phase of earthquake inception in Indonesia. The next significant signal rising of SEMF~CO4 was observed in 2008; the maximum value was 337 mV on March 04, 2008, i.e., about two months directly before the rapid phase of earthquake inception in China (May 12, 2008, M7.9).

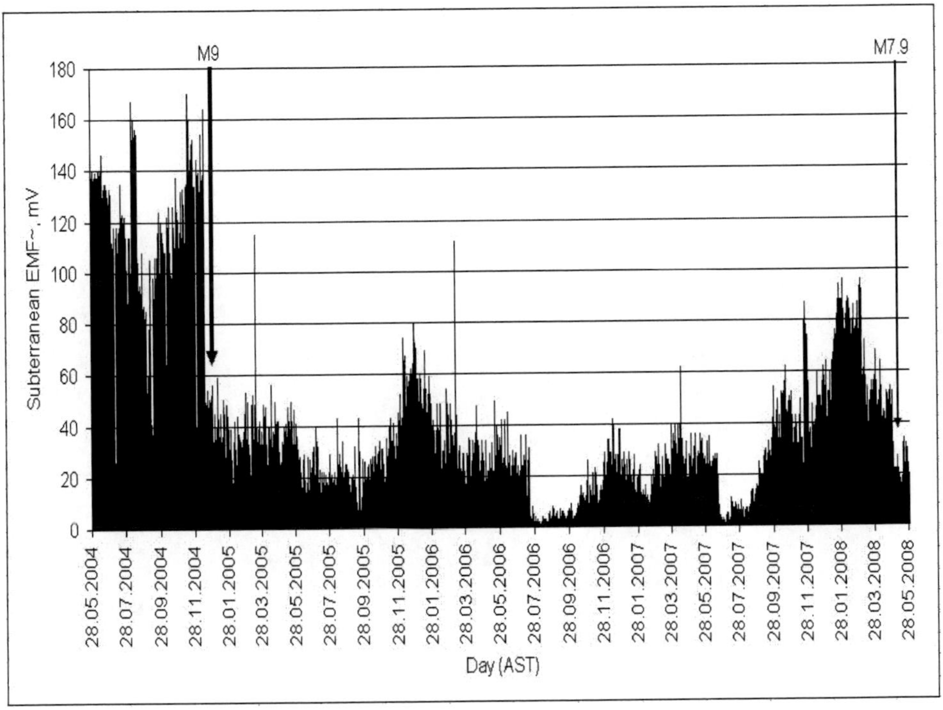

Figure 37. Column diagram of the largest daily signals of SEMF~ being taken in CO4~ channel over the interval May 28, 2004 to May 31, 2008. AST – apparent solar time in Petropavlovsk-Kamchatskiy, EMF~ – electromotive force of alternating currents, M9 – Indonesian earthquake magnitude (December 26, 2004), M7.9 – Chinese earthquake magnitude (May 12, 2008).

Figure 38 is a column diagram of the largest daily signals of SEMF~ taken in the CO4~ channel over the interval April 14 to May 20, 2008. Most of the fluctuations in the April indices are typical and can be considered normal. However, an anomalous main nonstationary process began to appear in the data for the SEMF~CO4 and it abruptly dropped down around May 1. It was not until the jump downward that the catastrophic earthquake with magnitude M7.9 in China (May 12, 2008) occurred, after about 12 days.

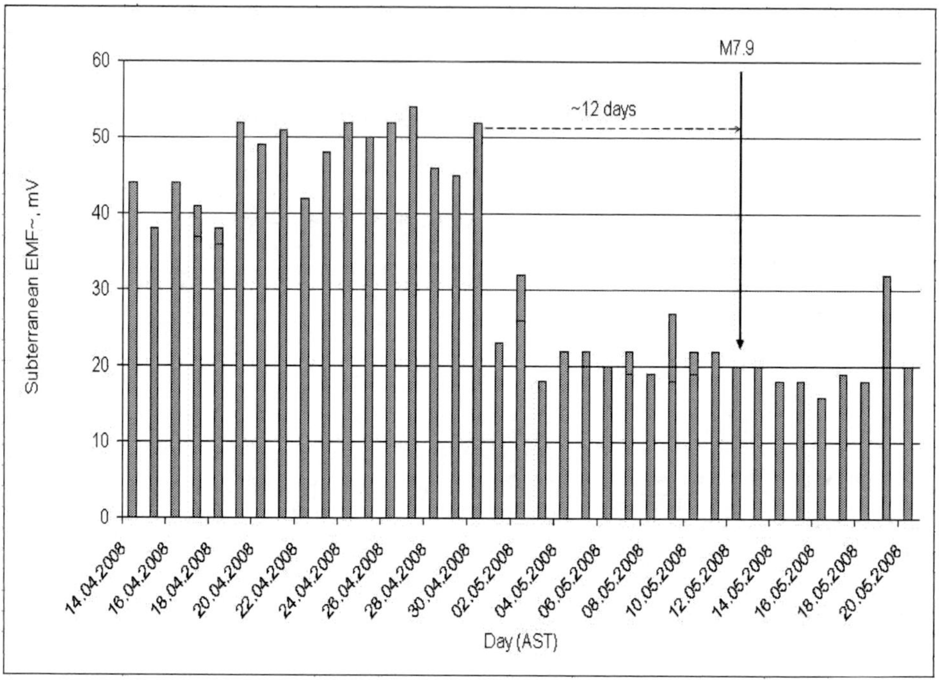

Figure 38. Column diagram of the largest daily signals of SEMF~ being taken in CO4~ channel over the interval April 14 to May 20, 2008. AST – apparent solar time in Petropavlovsk-Kamchatskiy, EMF~ – electromotive force of alternating currents, M7.9 – Chinese earthquake magnitude (May 12, 2008), ~12 days – days interval of PE signals abrupt jump down before earthquake M7.9.

Moving to the catastrophic China earthquake (May 12, 2008, M7.9), Figure 39 illustrates a column diagram of the largest daily signals of SEMF~ taken in the CO5~ channel over the interval May 28, 2004 to May 31, 2008. The column diagram shows large signals of SEMF~CO5 during more than six months directly before the tsunamigenic Indonesia earthquake (December 26, 2004, M9). The maximum value was 520 mV on November 17, 2004, i.e., a month before the rapid phase of earthquake inception in Indonesia. The next significant signal rising of SEMF~CO5 was observed in 2008; the maximum value was 294 mV on March 4, 2008, i.e., about two months directly before the rapid phase of earthquake inception in China (May 12, 2008, M7.9).

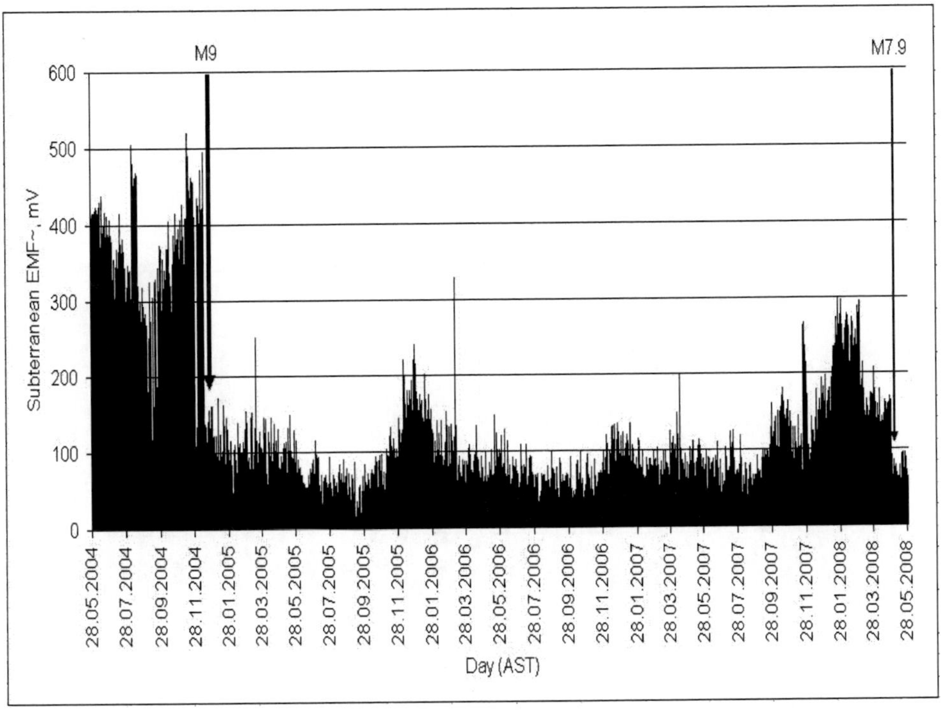

Figure 39. Column diagram of the largest daily signals of SEMF~ being taken in CO5~ channel over the interval May 28, 2004 to May 31, 2008. AST – apparent solar time in Petropavlovsk-Kamchatskiy, EMF~ – electromotive force of alternating currents, M9 – Indonesian earthquake magnitude (December 26, 2004), M7.9 – Chinese earthquake magnitude (May 12, 2008).

Figure 40 shows a column diagram of the largest daily signals of SEMF~ being taken in CO5~ channel over the interval April 14 to May 20, 2008. Most of the fluctuations in the April indices are typical and can be considered normal. However, an anomalous main nonstationary process began to appear in the data for the SEMF~CO5 and it abruptly dropped down around May 1. It was not until the jump downward that the catastrophic earthquake with magnitude M7.9 in China (May 12, 2008) occurred, after about 12 days.

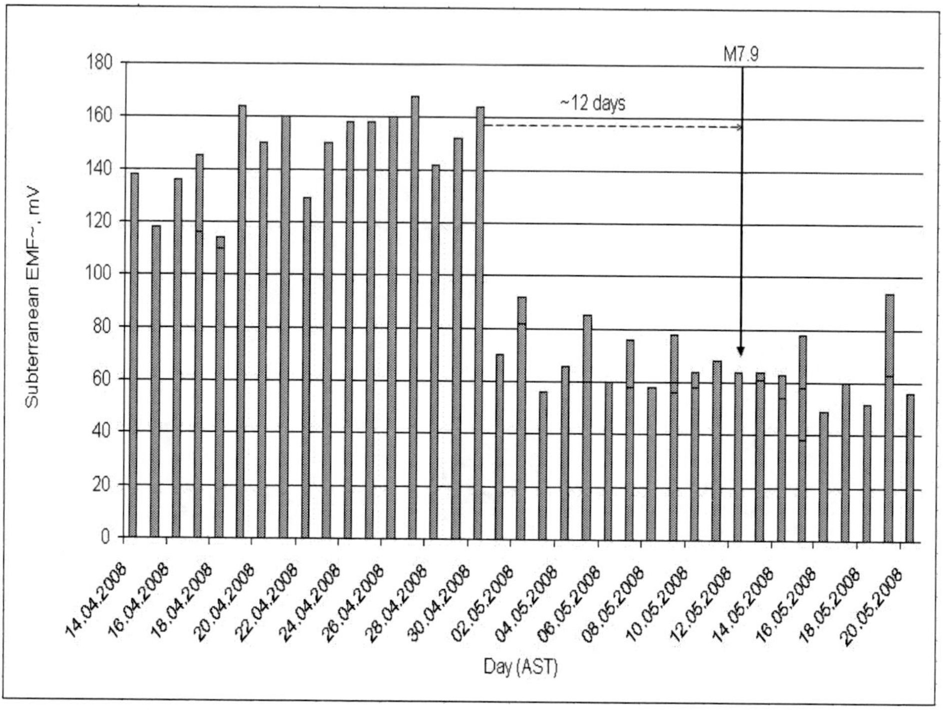

Figure 40. Column diagram of the largest daily signals of SEMF~ being taken in CO5~ channel over the interval April 14 to May 20, 2008. AST – apparent solar time in Petropavlovsk-Kamchatskiy, EMF~ – electromotive force of alternating currents, M7.9 – Chinese earthquake magnitude (May 12, 2008), ~12 days – days interval of PE signals abrupt jump down before earthquake M7.9.

Moving to the catastrophic China earthquake (May 12, 2008, M7.9), Figure 41 is a column diagram of the largest daily signals of SEMF~ being taken in CO6~ channel over the interval May 28, 2004 to May 31, 2008. The column diagram shows large signals of SEMF~CO6 for more than six months directly before the tsunamigenic Indonesia earthquake (December 26, 2004, M9). The maximum value was 511 mV on November 17, 2004, i.e., a month before the rapid phase of earthquake inception in Indonesia. The next significant signal rising of SEMF~CO6 was observed in 2008; the maximum value was 289 mV on March 4, 2008, i.e., about two months directly before the rapid phase of earthquake inception in China (May 12, 2008, M7.9).

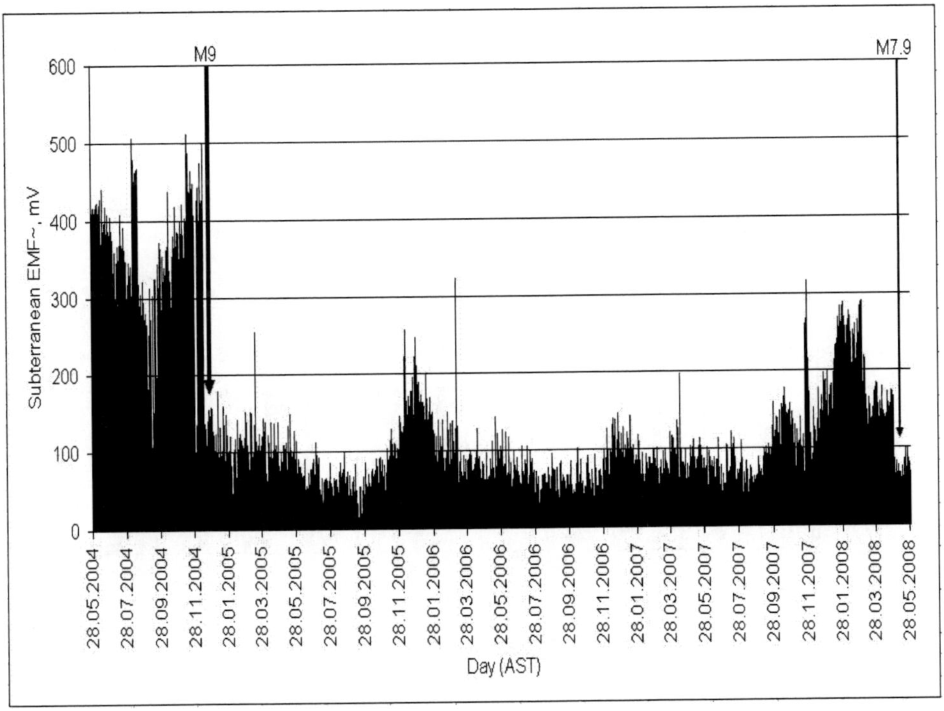

Figure 41. Column diagram of the largest daily signals of SEMF~ being taken in CO6~ channel over the interval May 28, 2004 to May 31, 2008. AST – apparent solar time in Petropavlovsk-Kamchatskiy, EMF~ – electromotive force of alternating currents, M9 – Indonesian earthquake magnitude (December 26, 2004), M7.9 – Chinese earthquake magnitude (May 12, 2008).

Figure 42 shows a column diagram of the largest daily signals of SEMF~ being taken in CO6~ channel over the interval April 14 to May 20, 2008. Most of the fluctuations in the April indices are typical and can be considered normal. However, an anomalous main nonstationary process began to appear in the data for the SEMF~CO6 and it abruptly dropped down around May 1. It was not until the jump downward that the catastrophic earthquake with magnitude M7.9 in China (May 12, 2008) occurred, after about 12 days.

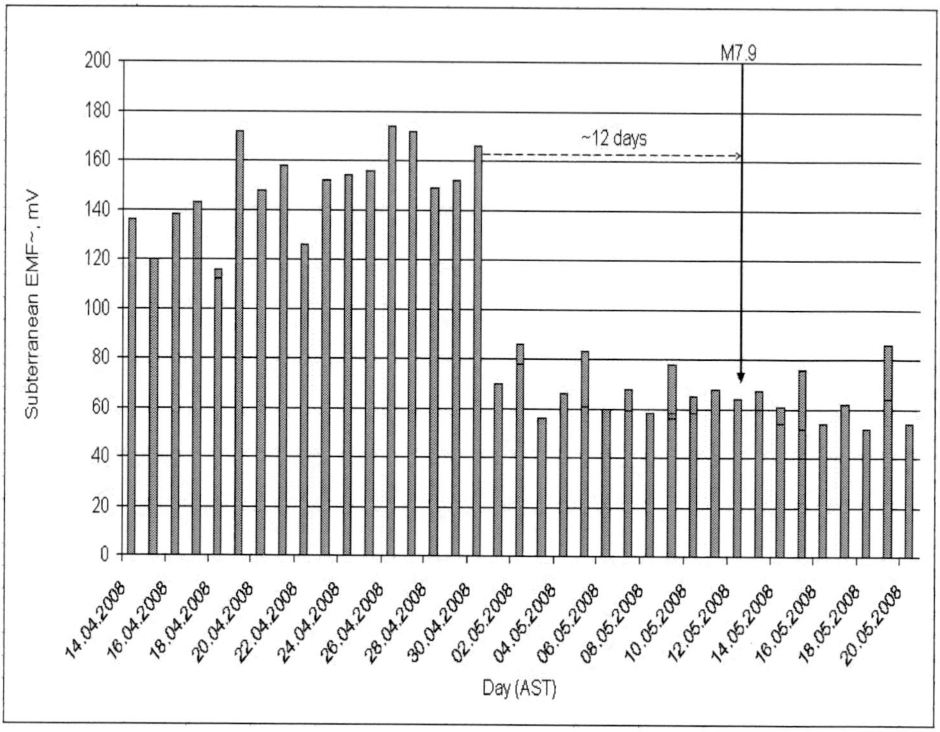

Figure 42. Column diagram of the largest daily signals of SEMF~ being taken in CO6~ channel over the interval April 14 to May 20, 2008. AST – apparent solar time in Petropavlovsk-Kamchatskiy, EMF~ – electromotive force of alternating currents, M7.9 – Chinese earthquake magnitude (May 12, 2008), ~12 days – days interval of PE signals abrupt jump down before earthquake M7.9.

Moving to the catastrophic China earthquake (May 12, 2008, M7.9), Figure 43 illustrates a column diagram of the largest daily signals of SEMF~ being taken in NE2C3~ channel over the interval May 28, 2004 to May 31, 2008. The column diagram shows large signals of SEMF~NE2C3 for more than six months directly before the tsunamigenic Indonesia earthquake (December 26, 2004, M9). The maximum value was 205 mV on November 25, 2004, i.e., about 20 days before the rapid phase of earthquake inception in Indonesia. The next significant signal rising of SEMF~NE2C3 was observed in 2008, the maximum value was 224 mV on 2008 March 2, 2008, i.e., about two months directly before the speedy phase of earthquake inception in China (May 12, 2008, M7.9).

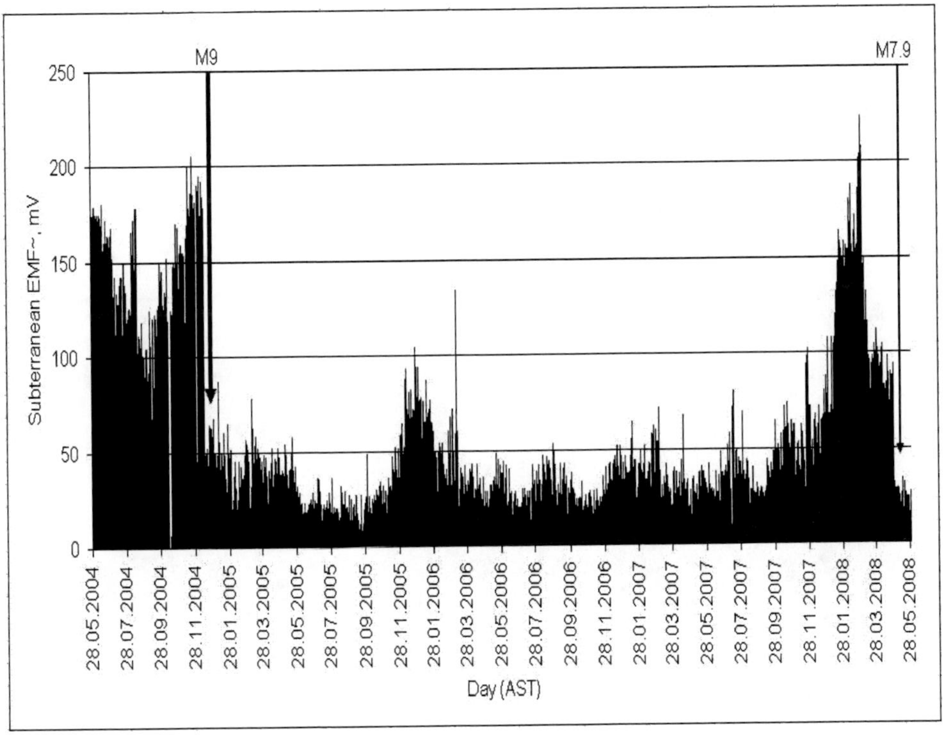

Figure 43. Column diagram of the largest daily signals of SEMF~ being taken in NE2C3~ channel over the interval May 28, 2004 to May 31, 2008. AST – apparent solar time in Petropavlovsk-Kamchatskiy, EMF~ – electromotive force of alternating currents, M9 – Indonesian earthquake magnitude (December 26, 2004), M7.9 – Chinese earthquake magnitude (May 12, 2008).

Figure 44 shows a column diagram of the largest daily signals of SEMF~ being taken in NE2C3~ channel over the interval April 14 to May 20, 2008. Most of the fluctuations in the April indices are typical and can be considered normal. However, an anomalous main nonstationary process began to appear in the data for the SEMF~NE2C3, and it abruptly dropped down around May 1. It was not until the jump downward that the catastrophic earthquake with magnitude M7.9 in China (May 12, 2008) occurred, after about 12 days.

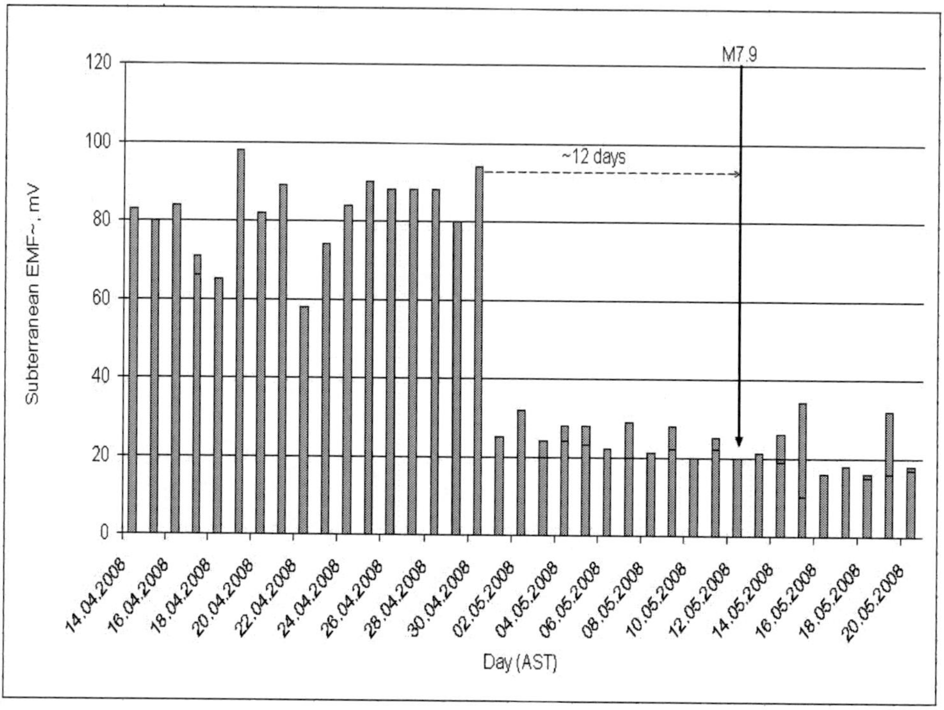

Figure 44. Column diagram of the largest daily signals of SEMF~ being taken in NE2C3~ channel over the interval April 14 to May 20, 2008. AST – apparent solar time in Petropavlovsk-Kamchatskiy, EMF~ – electromotive force of alternating currents, M7.9 – Chinese earthquake magnitude (May 12, 2008), ~12 days – days interval of PE signals abrupt jump down before earthquake M7.9.

Moving to the catastrophic China earthquake (May 12, 2008, M7.9), Figure 45 is a column diagram of the largest daily signals of SEMF~ taken in NE2C4~ channel over the interval May 28, 2004 to May 31, 2008. The column diagram shows large signals of SEMF~NE2C4 for more than six months directly before the tsunamigenic Indonesia earthquake (December 26, 2004, M9). The maximum value was 186 mV on 2004 November 25, i.e., about 20 days before the rapid phase of earthquake inception in Indonesia. The next significant signal rising of SEMF~NE2C4 was observed in 2008; the maximum value was 224 mV on March 2, 2008, i.e., about two months directly before the rapid phase of earthquake inception in China (May 12, 2008, M7.9).

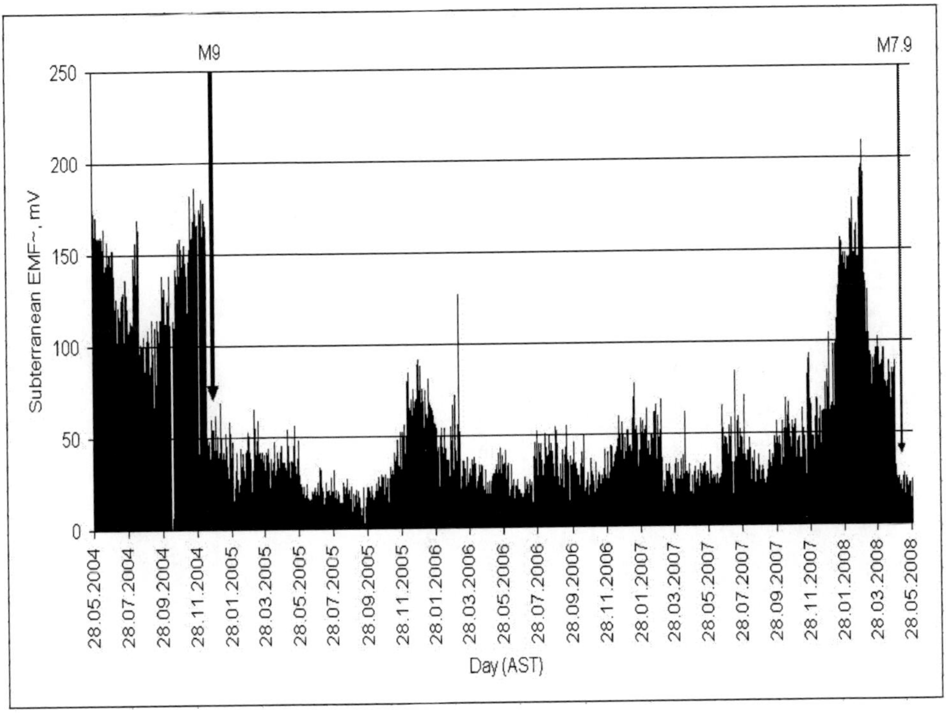

Figure 45. Column diagram of the largest daily signals of SEMF~ being taken in NE2C4~ channel over the interval May 28, 2004 to May 31, 2008. AST – apparent solar time in Petropavlovsk-Kamchatskiy, EMF~ – electromotive force of alternating currents, M9 – Indonesian earthquake magnitude (December 26, 2004), M7.9 – Chinese earthquake magnitude (May 12, 2008).

Figure 46 is a column diagram of the largest daily signals of SEMF~ being taken in the NE2C4~ channel over the interval April 14 to May 20, 2008. Most of the fluctuations in the April indices are typical and can be considered normal. However, an anomalous main nonstationary process began to appear in the data for the SEMF~NE2C4 and it abruptly dropped down around May 1. It was not until the jump downward that the catastrophic earthquake with magnitude M7.9 in China (May 12, 2008) occurred, after about 12 days.

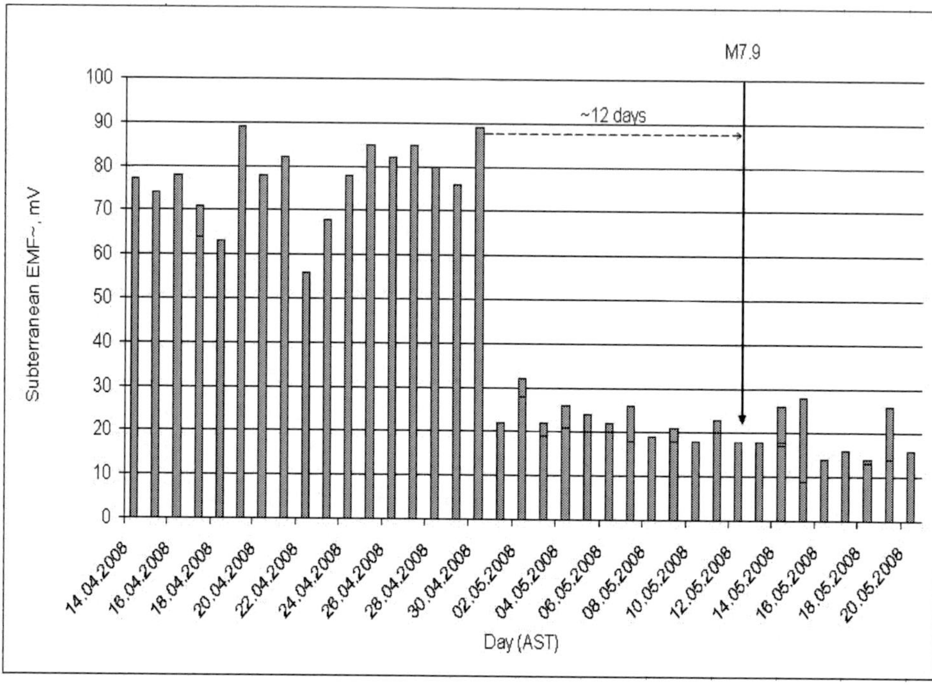

Figure 46. Column diagram of the largest daily signals of SEMF~ being taken in NE2C4~ channel over the interval April 14 to May 20, 2008. AST – apparent solar time in Petropavlovsk-Kamchatskiy, EMF~ – electromotive force of alternating currents, M7.9 – Chinese earthquake magnitude (May 12, 2008), ~12 days – days interval of PE signals abrupt jump down before earthquake M7.9.

DISCUSSION

Figures 1–24 show that 24 subterranean PE parameters of alternating currents out of 29 responded to a rapid onset of tsunamigenic earthquake with magnitude M9 on December 26, 2004 (Indonesia).

Figure 26, Figure 28, Figure 30, Figure 32, Figure 34, Figure 36, Figure 38, Figure 40, Figure 42, Figure 44 and Figure 46 show that 11 subterranean PE parameters of alternating currents out of 29 responded to a rapid onset of damaging earthquake with magnitude M7.9 on May 12, 2008 (China).

Thus, a magnitude difference of M1.1 in the earthquakes resulted in a greater response in the 13 subterranean PE parameters in the first case than in the second one.

Reviews of Figures 1–46 show contrasting differences observed during jumps downward in Figures 1–24 incomparison with Figure 26, Figure 28, Figure 30, Figure 32, Figure 34, Figure 36, Figure 38, Figure 40, Figure 42, Figure 44 and Figure 46.

As illustrated in Figures 1–24, contrast calculations are carried out according to the 11 largest values before jumps downward and according to the 11 largest values after jumps downward. Table 1 lists the contrast values of the first case (Indonesia earthquake).

Table 1.

Figure	Channel	Contrast
1	Subterranean EMF ~C23	3.6
2	Subterranean EMF ~C45	4.3
3	Subterranean EMF ~CO1	3.5
4	Subterranean EMF ~CO2	3.4
5	Subterranean EMF ~CO3	3.0
6	Subterranean EMF ~CO4	3.1
7	Subterranean EMF ~CO5	3.3
8	Subterranean EMF ~CO6	3.3
9	Subterranean EMF ~NE2C1	3.1
10	Subterranean EMF ~NE2C2	3.0
11	Subterranean EMF ~NE2C3	3.5
12	Subterranean EMF ~NE2C4	3.5
13	Subterranean EMF ~NE2C5	3.1
14	Subterranean EMF ~NE2C6	3.1
15	Subterranean EMF ~SW34	3.9
16	Subterranean EMF ~SW45	3.0
17	Subterranean EMF ~SW56	5.4
18	Subterranean EMF ~SWO1	3.4
19	Subterranean EMF ~SWO2	3.3
20	Subterranean EMF ~SWO3	3.2
21	Subterranean EMF ~SWO4	3.3
22	Subterranean EMF ~SWO5	3.5
23	Subterranean EMF ~SWO5	3.3
24	Subterranean EMF ~NE	3.4
		Mean contrast 3.4

As illustrated in Figure 26, Figure 28, Figure 30, Figure 32, Figure 34, Figure 36, Figure 38, Figure 40, Figure 42, Figure 44 and Figure 46, contrast calculations are carried out according to the 12 largest values before jumps downward and according to the 12 largest values after jumps downward. Table 2 lists contrast values of the second case (China earthquake).

Table 2.

Figure	Channel	Contrast
26	Subterranean EMF ~NE	2.4
28	Subterranean EMF ~C23	2.2
30	Subterranean EMF ~C45	2.1
32	Subterranean EMF ~CO1	2.2
34	Subterranean EMF ~CO2	2.3
36	Subterranean EMF ~CO3	2.2
38	Subterranean EMF ~CO4	2.2
40	Subterranean EMF ~CO5	2.2
42	Subterranean EMF ~CO6	2.3
44	Subterranean EMF ~NE2C3	3.4
46	Subterranean EMF ~NE2C4	3.5
		Mean contrast 2.5

Comparison of mean contrasts for the first and second cases shows that rapid onset of the tsunamigenic earthquake with magnitude M9 on December 26, 2004 (Indonesia) required 1.4 times higher mean contrast than rapid onset of the damaging earthquake with magnitude M7.9 on May 12, 2008 (China).

Figure 25, Figure 27, Figure 29, Figure 31, Figure 33, Figure 35, Figure 37, Figure 39, Figure 41, Figure 43 and Figure 45 illustrate the results of subterranean PE measurements over the interval May 28, 2004 to May 31, 2008. After reviewing the figures, we found abrupt changes of subterranean PE signals before damaging earthquakes in Indonesia (December 26, 2004) and China (May 12, 2008). Availability of similar subterranean PE signals demonstrates a nonrandom phenomenon pattern. Comparison of both phases of rapid onset of damaging earthquakes demonstrates comparatively equal time frames (difference of one day) of abrupt jumps of subterranean PE signals and earthquakes' onset. With the Indonesian earthquake, this interval was about 11 days between signal jump and earthquake point. With the Chinese earthquake, this interval was about 12 days between signal jump and earthquake point. Such a closeness in intervals shown in this chapter is evidence of the regularity of the phenomenon being observed.

Thus, the relationship between subterranean PE measurements with Kamchatka and damaging earthquakes in Indonesia and China is demonstrated. The earthquakes were rather different by location. One was oceanic and another was continental. But, even though similar subterranean PE signals in the rapid onset stage were being observed, there were abrupt changes between subterranean PE signals and earthquakes points.

The work of Fraser-Smith et al. (1990) confirms evidence of the proton tectogenesis model (Larin, 1980; Larin, 1993). Figure 3 in Fraser-Smith et al. (1990) clearly shows jumps in ULF-measurements (from background level of about 1 nT/√Hz to about 22 nT/√Hz), which marked the beginning of the rapid Loma Prieta earthquake inception stage about 13 days before start-up. ULF-signal jump (from about 5 nT/√Hz to about 58 nT/√Hz) occurred three hours before start-up. In the framework of proton tectogenesis (Larin, 1980; Larin, 1993), a 13-days interval of rapid earthquake inception corresponds to half a solar day (solar day, the time of complete pivot turn of the solar photosphere, equal to about 27 earth days). The phase of rapid onset of the Loma Prieta earthquake was begun with a ULF-signal jump at the Corralitos station about 13 days before earthquake start-up. The jump contrast of the ULF-signal was about (~22 nT/~1 nT)≈22. The phase of start-up of the Loma Prieta earthquake was begun with a ULF-signal jump at the same station about three hours before the earthquake's onset; the jump contrast was (~58 nT/~5 nT)≈11.6, which was about half of the jump contrast in the beginning of rapid phase of the Loma Prieta earthquake inception. Thus, the jump contrast of the ULF-signal being observed half a solar day before earthquake was two times higher than the jump contrast of ULF-signal being observed three hours before the Loma Prieta earthquake.

Electromagnetic interpretation results of ULF magnetic field measurements under consideration of the inherent difficulties showed that direct measurements of subterranean EMF in the framework of current networks give more appropriate precursor information of rapid earthquake onset. In perspective, subterranean PE measurements at many stations will make it possible to use a search method of electrical failure of compound electrical networks for operative forecasting of earthquakes. As it is known, passive test systems are used to repair compound electric devices, and to detect both critical failures (breaks and short circuits) and noncritical process variations of electrical subunits of the testing device. Thus, the basis of a subterranean PE system for engineering an expert system of operative forecasting of earthquakes can be described by similar contexts which are being used with passive test systems engineering.

Conclusions

As a result of the analysis of the relationship of the contrast in subterranean PE signals, the power of being prepared for earthquakes has been demonstrated.

It was found that the similarily in the behavior of subterranean PE signals provides a warning of the onset of an earthquake. It should be noted that subterranean PE signals' waveform and amplitude have direct relevance to the power and distance of the onset of an earthquake.

A time interval was found in the rapid phase of earthquake inception after abrupt changes of subterranean PE signals. It should be noted that distance and power had a weak influence on time interval of the rapid phase of earthquake inception. The time interval was 11 days in the Indonesian case and 12 days in the Chinese case.

The correlation of subterranean PE signals' waveform and amplitude with the power and distance of the earthquake focus is evidence of the electrical current network nature of the rapid phase of earthquake inception.

The results of the Kamchatkian subterranean PE measurements support the effectiveness of using the concept of proton tectogenesis (Larin, 1980; Larin, 1993) to interpret the rapid phase of the onset of a catastrophic earthquake.

The results of this study indicate that it is necessary to place such subterranean PE stations in other dangerous seismic regions of the Pacific Ring, the Indian Ocean and other such regions of the Earth.

Acknowledgments

I thank D. Kuznetsov for his assistance and comments.

References

Alefeld, G.; Vökl, J. *Hydrogen in Metals*; Springer: Berlin, GE, 1978; Vol. 1, 2.

Alfven, H. *Cosmic plasma*; Reidel: Dordrecht, NL, 1981.

Båth, M. *Scientia*. 1966, an. LX, CI, N., ser.VII, 1-10.

Bobrovskiy, V. S. (2005). Cosmo-meteo-seismic conception of Gutenberg. On appropriateness prolongation of the forecast range of strong earthquake in winter-spring'2006 on Kamchatka. *Abstract Journal Geophysics*. All-Russian institute for Scientific and Technical Information Russian Academy of Science. 1681-V2005. dep@viniti.ru to copy manuscript.

Fraser-Smith, A. C.; Bernardi, A.; McGill, P. R.; et al. *Geophysical Research Letters*. 1990, *vol. 17*, 1465-1468.

Gutenberg, B. *Grundlagen der Erdbebenkunde*; Verlag von Gebrüder Borntraeger: Berlin, GE, 1927.

Hattori, K.; Takahashi, I.; Yoshino, C.; Isezaki, N.; et al. *Phys. Chem. Earth*. 2004, *vol. 29*, 481-494.

Hayakawa, M.; Hattori, K.; Ohta, K. *Sensors*. 2007, *vol. 7*, 1108-1122.

Khusnatdinov, N. N.; Petrenko, V. F.; Zaretskii, A.V. *J. Phys. Colloques*. 1987, *vol. 48*, 105-108.

Larin, V. N. *Hypothesis of primordially hydride Earth*; Nedra: Moscow, RU, 1980, p. 216.

Larin, V. N. In *Hydridic Earth: the New Geology of Our Primordially Hydrogen-Rich Planet*; Hunt, C. W.; Ed.; Polar Publishing: Calgary, CA, 1993.

Uyeda, S.; Hayakawa, V.; Nagao, T.; et al. *Proc. Nation. Acad. Sci. USA*. 2002, *vol. 99*, 7352-7355.

Varotsos, P.; Alexopoulos, K. *Tectonophysics*. 1984, *vol. 110*, 73-98.

Varotsos, P.; Sarlis, N.; Skordas, E.; Lazaridou, M. *Tectonophysics*. 2006, *vol. 412*, 279-288.

Zaretskiy, A. V.; Petrenko, V. F.; Chesnakov, V. A. *Phys. Stat. Sol. (a)*. 1988, *vol. 109*, 373-381.

INDEX

A

B

C

D

H

I

J

K

L

M

N

S

X

Y